LIFE MOVEMENTS IN PLANTS

LIFE MOVEMENTS IN PLANTS

VOL. 2

J. C. BOSE

ANMOL PUBLICATIONS
New Delhi (India)

ANMOL PUBLICATIONS
4378/4B, Ansari Road
Daryaganj, New Delhi-110 002

Reprint 1989
Reprinted 1993

ISBN 81-7041-172-6 (Set)
ISBN 81-7041-174-2 (Vol. 2)

PRINTED IN INDIA
Published by J.L. Kumar for Anmol Publications, New Delhi-110 002 and
Printed at Mehra Offset Press, Delhi-110 006.

PREFACE TO VOLUME II.

I HAVE in the present volume dealt with the intricate phenomena of different tropisms. The movements in plants under the stimuli of the environment—the twining of tendrils, the effect of temperature, the action of light inducing movements sometimes towards and at other times away from the stimulus, the diametrically opposite responses of the shoot and the root to the same stimulus of gravity, the day and night positions of organs of plants—these, and many others present such diversities that it must have appeared a hopeless endeavour to discover any fundamental reaction applicable in all cases. It has therefore been customary to assume different sensibilities especially evolved for the advantage of the plant. But teleological argument and the use of descriptive phrases, like positive and negative tropism, offer no real explanation of the phenomena. Thus to quote Pfeffer "When we say that an organ curves towards a source of illumination, because of its heliotropic irritability we are simply expressing an ascertained fact in a conveniently abbreviated form, without explaining why such curvature is possible or how it is produced. . . . Many observers have unfortunately devoted their attention to artificially classifying the phenomenon observed, and have entirely neglected the explanation of causes underlying them." He also adds that in regard to the phenomenon of growth and its variations, an empirical treatment is all that is possible in the present state of our knowledge; but deduction from results of experimental investigation "still remains the ideal of physiology, and only when this ideal has been attained, shall we be able to obtain a comprehensive view of the interacting factors at work in the living organism."

In my previous work on "Plant Response" (1906) I described detailed investigations on irritability of plants which I carried out with highly sensitive recorders. The plant was thus made to tell its own story by means of its self-made records. The results showed that there is no specific difference in physiological reaction of different organs to justify the assumption of positive and negative irritabilities. A generalisation was obtained which gave a complete explanation of diverse movements in plants. The results were fully confirmed by an independent method of inquiry, namely that of electric response, which I have been able to elaborate so as to become a very important means of research.

The investigations described in the present volume not only support the conclusions reached in my earlier works, but have led to important additions. It is evident that the range of our investigation is limited only by our power of recording the rate of plant-movement, that is to say, in the measurement of length and time. In these respects the instruments that I have been able to devise have surpassed my sanguine expectations. The Resonant Recorder traces time-intervals as short as a thousandth part of a second, while my Balanced Crescograph enables us to measure variation of rate of growth as minute as $\frac{1}{1000}$ millionth of an inch per second, the sensitiveness of this apparatus thus rivals that of the spectroscope. The increasing refinement in our experimental methods cannot but lead to important advances towards a deeper understanding of underlying reactions in the living organism.

I shall here draw attention to only a few of the important results given in the present volume. The tropic effect of light has been shown to have a definite relation to the quantity of incident light. A complete tropic curve has been obtained from sub-minimal to maximal stimulation which shows the inadequacy of Weber's law, for the sub-minimal stimulus induces a *qualitative* difference in physiological reaction. It has further been shown that the prevalent idea that perception and heliotropic excitation are two distinct phenomena is without any foundation.

With reference to the effect of ether waves on plants, I have given an account of my discovery of the response of all plants to wireless stimulation, the results being similar to that induced by visible light. The perceptive range of the plant is thus infinitely greater than ours; for it not only perceives, but also responds to different rays of the vast ethereal spectrum.

The results obtained by the method of geo-electric response show that the responsive reaction of the root is in no way different from that of the shoot, the opposite movements being due to the fact that in the shoot the stimulation is direct, and in the root it is indirect.

Full description is given of the new method of physiological exploration by means of the electric probe, by which the particular layer which perceives the stimulus of gravity is definitely localised. The method of electric probe is also found to be of extended application in the detection of physiological changes in the interior of an organ.

An important factor of nyctitropic movements, hitherto unsuspected. is the effect of variation of temperature on geotropic curvature. This and other co-operative factors have been fully analysed, and a satisfactory explanation has been offered of various types of diurnal movement.

A generalisation has been obtained which explains all the diverse movements of plants, under all modes of stimulation: *it has been shown that direct stimulation induces contraction and retardation of growth, and that indirect stimulation induces an expansion and acceleration of growth.*

Another generalisation of still greater importance is the establishment of identical nature of physiological reaction in the plant and the animal, leading to advances in general physiology. Thus the discovery of a method for immediate enhancement or inhibition of nervous impulse in the plant led to my success in the control of nervous impulse in the animal. Another important discovery was the dual nervous impulses in plants, and I have very recently been able to establish, that the nervous impulse generated in the animal nerve by stimulus is not single, but double.

The study of the responsive phenomena in plants must thus form an integral part of physiological investigation into various problems relating to the irritability of all living tissues, and without such study the investigation must in future remain incomplete.

October 1919. J. C. BOSE.

CONTENTS.

PART III.

TROPISM IN PLANTS.

XXII.—THE BALANCED CRESCOGRAPH.

PAGE.

Principle of the Method of Balance—Compensating movement—Growth-scale—Sensitiveness of the Crescographic Balance—Effect of CO_2—Effect of anæsthetics . . 255

XXIII.—ON TROPIC MOVEMENTS.

Complexity of the problem—Contradictory nature of responses—Two classes of tropic responses—Longitudinal transmission of effect of stimulus—Transverse transmission of effect of stimulus—Modification of tropic curvature by conducting power of tissues and differential excitability of the organ 268

XXIV.—TROPIC CURVATURE WITH LONGITUDINAL TRANSMISSION OF EFFECT OF STIMULUS.

Dual impulses, *positive* and *negative*, caused by stimulus—Direct and Indirect stimulus—Tropic effect of Indirect stimulation 271

XXV.—TROPIC CURVATURE WITH TRANSVERSE TRANSMISSION OF EFFECT OF STIMULUS.

PAGE.

Turgor variation under transverse transmission of stimulus-effect—Tropic responses of pulvinated and growing organs to unilateral ttimulation—Direct unilateral stimulation—Indirect unilateral stimulation—Difference of effects induced by Direct and Indirect stimulation—Laws of tropic curvature 279

XXVI.—MECHANOTROPISM : TWINING OF TENDRILS.

Anomalies of mechanotropism—Effects of indirect and direct electric stimulation on growth of tendril—Effect of direct and indirect mechanical stimulus—Immediate and after-effect of stimulus—Inhibitory action of stimulus—Response of less excitable side of the tendril—Relative intensities of responses of upper and under sides of tendril of *Passiflora*—Negative curvature of tendril 288

XXVII.—ON GALVANOTROPISM.

Polar effects of electric current on growth—Effect of anode and cathode on growth 301

XXVIII.—ON THERMONASTIC PHENOMENA.

Effect of temperature—Different thermonastic organs—Two types of response : Positive and Negative—Effect of rise of temperature and of stimulus on thermonastic organs—Law of thermonastic reaction 305

XXIX.—ON PHOTOTROPISM.

PAGE.

Complexity of problem of phototropic reaction—Action of light—Positive phototropic curvature of pulvinated organs—Positive phototropic curvature of growing organs—Phenomenon of recovery—Immediate and after-effect of light on growth—Latent period of phototropic reaction—Growth variation induced by flash of light from a single spark—Maximum positive curvature under continued action of light 313

XXX.—DIA-PHOTOTROPISM AND NEGATIVE PHOTOTROPISM.

Differential excitability of two halves of pulvinus of *Mimosa* —Transformation of positive to negative curvature—Tropic effect under sunlight—Negative phototropism of shoot and root 328

XXXI.—RELATION BETWEEN THE QUANTITY OF LIGHT AND THE INDUCED PHOTOTROPIC CURVATURE.

Effect of increasing intensity of light on pulvinated and growing organs—Effect of increasing angle—Effect of duration of exposure 338

XXXII.—THE PHOTOTROPIC CURVE AND ITS CHARACTERISTICS.

Summation of stimulus—General consideration—The general characteristic curve—Characteristics of simple phototropic curve—Variation of susceptibility for excitation in different parts of the curve—Effect of sub-minimal stimulus—The complete phototropic curves of pulvinated and growing organs—Limitation of Weber's law . 346

XXXIII.—TRANSMITTED EFFECT OF PHOTIC STIMULATION.

PAGE.

Effect of light applied on tip of *Setaria*—Response of growing region to unilateral stimulus—Effect of simultaneous stimulation of the tip and the hypocotyl—Algebraical summation of effects of direct and indirect stimuli 362

XXXIV.—ON PHOTONASTIC CURVATURES.

Phototropic response of anisotropic organs—Positive para-heliotropism—Negative para-heliotropism—Responses of pulvinated and growing organs to light . . 378

XXXV.—EFFECT OF TEMPERATURE ON PHOTOTROPIC CURVATURE.

Effect of temperature on excitability—Effect of temperature on conduction—Phototropic response of tendrils—Seasonal variation of phototropic curvature—Antagonistic effects of light and of rise of temperature . . 388

XXXVI.—ON PHOTOTROPIC TORSION.

Torsional response to light—Effect of different modes of lateral stimulation—Effect of differential excitability on the direction of torsion—Laws of torsional response—Complex torsion under light—Advantages of the Method of Torsional Response—The Torsional Balance—Determination of the direction of stimulus . 397

XXXVII.—RADIO-THERMOTROPISM.

Effect of infra-red radiation—Positive radio-thermotropism—Dia-radio-thermotropism—Negative radio-thermotropism 410

XXXVIII.—RESPONSE OF PLANTS TO WIRELESS STIMULATION.

PAGE.

Effects of different rays of spectrum on growth—The wireless system—Mechanical and electrical responses of *Mimosa* to Hertzian waves—Effect of wireless stimulation on growth of plants 416

XXXIX.—GEOTROPISM.

Direction of the stimulus of gravity—The Geotropic Recorder—Determination of the character of geotropic reaction—Theory of statoliths—Determination of the latent period—The complete geotropic curve—Determination of effective direction of stimulus of gravity—Algebraical summation of effects of geotropic and photic stimulus—Analogy between the effects of stimulus of light and of gravity—Relation between the directive angle and geotropic reaction—Differential geotropic excitability 425

XL.—GEO-ELECTRIC RESPONSE OF SHOOT.

Electric response to direct and indirect stimulation—Experimental arrangement for obtaining geo-electric response—Geo-electric response of the upper and lower sides of the organ—Method of Axial Rotation—Characteristics of geo-electric response—Physiological character of geo-electric response—Effect of differential excitability of the organ—Law determining the relation between angle of inclination and geotropic effect—Method of Vertical Rotation—Electric response through an entire cycle—Relation between angle of vertical rotation and intensity of geo-tropic reaction 442

XLI.—MECHANICAL AND ELECTRICAL RESPONSE OF ROOT TO VARIOUS STIMULI.

PAGE.

Mechanical and electrical response to Direct stimulation—Mechanical and electrical response to Indirect stimulation—Effect of unilateral stimulation applied at the root-tip 461

XLII.—GEO-ELECTRIC RESPONSE OF ROOT.

Geo-electric response of the root-tip—Electric response in the growing region of root—Differential effect between the tip and the growing region—Geo-perception at the root-tip 467

XLIII.—LOCALISATION OF GEO-PERCEPTIVE LAYER BY MEANS OF THE ELECTRIC PROBE.

Principle of the method of electric exploration—The Electric Probe—Electric exploration of the geo-perceptive layer—Geo-electric reaction at different depths of tissues—Microscopical examination of the maximally excited layer—Influence of season on geo-electric response—Tests of insensitive specimens—Reaction at lower side of the organ—The Method of Transverse Perforation 478

XLIV.—ON GEOTROPIC TORSION.

Arrangement for torsional response—Algebraical summation of geotropic, and phototropic effects—Balance of geotropic by phototropic action—Comparative balancing effects of white and red lights—Effect of coal gas on photo-geotropic balance 503

XLV.—ON THERMO-GEOTROPISM.

PAGE.

Necessary conditions for geotropic curvature—Modifying influence of temperature on geotropic curvature—Magnetic analogue—Tropic equilibrium under varying intensities of stimulus—Effect of variation of temperature on geotropic torsion—Variation of apo-geotropic curvature under thermal change—Effect of variation of temperature on dia-geotropic equilibrium 509

PART IV.

NIGHT AND DAY MOVEMENTS IN PLANTS.

XLVI.—DIURNAL MOVEMENTS IN PLANTS.

Complexity of the problem—The different factors involved—Autonomous movements—Epinasty and hyponasty—Positive and negative thermonasty—Thermo-geotropism—Classification of diurnal movements—Discriminating tests for classification—Diurnal variation of light and of temperature 523

XLVII.—DIURNAL MOVEMENT DUE TO ALTERNATION OF LIGHT AND DARKNESS.

Experimental arrangements—The Quadruplex Nyctitropic Recorder—Diurnal movement of the leaflet of *Cassia alata*—Effect of variation of temperature—Effect of variation of light—Diurnal movement of the terminal leaflet of *Desmodium gyrans*—The 'midday sleep' . 535

XLVIII.—DIURNAL MOVEMENT DUE TO VARIATION OF TEMPERATURE AFFECTING GROWTH.

Tropic and nastic movements—Distinction between thermonastic and thermo-geotropic action—Diurnal movement of *Nymphæa*—Action of light—Effect of variation of temperature 546

XLIX.—DAILY MOVEMENT IN PLANTS DUE TO THERMO-GEOTROPISM.

PAGE.

Characteristics of thermo-geotropic movements—Diurnal movement of Palm trees—Diurnal movement of procumbent stems and of leaves—Continuous diurnal record for successive thermal noon—Modification of the diurnal curve—Effect of fluctuation of temperature—Effect of restricted pliability of the organ—Effect of age—Effect of season—Reversal of the normal rhythm—Effect of constant temperature—Diurnal movement in inverted position 554

L.—THE AFTER-EFFECT OF LIGHT.

Electric after-effect of light—After-effect at pre-maximum, maximum, and post-maximum—Tropic response under light, and its after-effects at pre-maximum, maximum, and post-maximum 569

LI.—THE DIURNAL MOVEMENT OF THE LEAF OF *MIMOSA*.

Four different phases in the diurnal record of *Mimosa*—Different factors determining its diurnal movement—Diurnal variation of geotropic torsion—Autonomous pulsation of the leaf of *Mimosa*—The Photometric Recorder—Effect of direct light—The evening spasmodic fall of the leaf—Diurnal movement of the amputated petiole—Diurnal curve of the petiole of *Cassia alata*—Response of *Mimosa* to darkness at different parts of the day—After-effect of light at pre-maximum, maximum, and post-maximum . . 576

ILLUSTRATIONS.

FIGURE. PAGE.

93. Arrangement for compensation of growth-movement by equal subsidence of plant-holder 257
94. Photograph of the Balanced Crescograph . . 258
95. Balanced Crescographic record 260
96. Record showing the effect of CO_2 265
97. Effect of ether and of chloroform . . . 266
98. Diagrammatic representation of effects of Indirect and Direct stimulation 275
99. Tropic curvature of *Crinum* 276
100. Turgor variation caused by Indirect stimulation 281
101. Response of *Mimosa* leaf under transverse transmission of effect of electric stimulus 282
102. Diagrammatic representation of Indirect and Direct stimulation of tendril 290
103. Record by Method of Balance 291
104. Variation of growth under direct stimulation . . 292
105. Positive curvature of tendril of *Cucurbita* . . . 295
106. Diagrammatic representation of effects of Indirect and Direct unilateral stimulation of tendril . . 296
107. Retardation of rate of growth under cathode . . 303
108. Acceleration of rate of growth under anode . . . 303
109. Thermonastic and radionastic responses of petal of *Zephyranthes* 308
110. The Thermonastic Recorder 309
111. Negative thermonastic response of *Nymphæa* . . 310
112. Successive positive responses of the terminal leaflet of bean plant 317
113. Positive response and recovery under moderate phototropic stimulation 318
114. Persistent positive curvature under stronger stimulation . 318

FIGURE.		PAGE.
115.	Immediate and after-effect of stimulus of light on growth	320
116.	Latent period for photic stimulation	324
117.	Effect of a single electric spark on growth	325
118.	Responses of *Mimosa* leaf to light from above	330
119.	Responses of *Mimosa* leaf to light from below	330
120.	Record of effect of continuous application of light on upper half of pulvinus of *Mimosa*	331
121.	Positive and negative phototropic response of *Oryza*	335
122.	Leaf of *Desmodium gyrans*	339
123.	The Oscillating Recorder	340
124.	Tropic effect of increasing intensity of light on the leaflet of *Desmodium gyrans*	341
125.	Tropic effect of increasing intensity of light on growing organ (*Crinum*)	341
126.	The Collimator	342
127.	Effect of angle of inclination of light on tropic curvature of pulvinated organ	343
128.	Effect of angle of inclination on growth-curvature	343
129.	Effect of increasing duration of exposure to light	344
130.	Effect of continuous electric and photic stimulation on rate of growth	348
131.	Characteristic curve of iron	351
132.	Simple characteristic curve of phototropic reaction	351
133.	Complete phototropic curve of pulvinated organ	358
134.	Complete phototropic curve of growing organ	359
135.	Arrangement for local application of light	357
136.	Response of seedling of *Setaria* to light	368
137.	Effect of application of light to the growing hypocotyl of *Setaria*	370
138.	Response to direct and indirect photic stimulus	373
139.	Diagrammatic representation of the effects of direct and indirect stimulation of *Setaria*	375
140.	Photonastic curvature of creeping stem of *Mimosa pudica*	380
141.	Positive phototropic response of *Erythrina indica*	382

FIGURE.		PAGE.
142.	Response of leaflet of *Mimosa* to light . . .	383
143.	Response of leaflet of *Averrhoa* to light . . .	383
144.	Diagrammatic representation of different types of phototropic response	384
145.	Phototropic curvature of tendril of *Passiflora* . .	392
146.	Effect of rise of temperature on phototropic curvature	394
147.	After-effect of rise of temperature	395
148.	Arrangement for record of torsional response . .	399
149.	Record of torsional response of pulvinus of *Mimosa pudica*	400
150.	Leaflets of *Cassia alata*	404
151.	Positive response to thermal radiation	413
152	Record of positive, neutral, and reversed negative curvature under thermal radiation	414
153.	Diagrammatic representation of the wireless system .	419
154.	Mechanical response of *Mimosa* leaf to electric waves .	420
155.	Electric response of *Mimosa* to Hertzian wave . .	420
156.	Record of responses of growing organs to wireless stimulation	422
157.	The Quadruplex Geotropic Recorder	428
158.	Effect of alternate application of cold on upper and lower sides of the organ	430
159.	Geotropic response of flower stalk of Tube-rose .	433
160.	Geotropic response of *Tropæolum*	433
161.	The Complete Geotropic Curve	435
162.	Diagrammatic representation of photic and geotropic stimulation	436
163.	The effect of super-imposition of photic stimulus .	436
164.	Diagrammatic representation of the mechanical and electrical response	443
165.	Diagrammatic representation of geo-electric response .	447
166.	Diagrammatic representation of Methods of Axial and Vertical Rotation	449
167.	Diagrammatic representation of the geo-electric response of the shoot	450
168.	Geo-electric response of the petiole of *Tropæolum* .	452

FIGURE.		PAGE.
169.	Geo-electric response of the scape of *Uriclis* . .	453
170.	Mechanical and electric response to indirect stimulation	463
171.	Diagrammatic representation of mechanical and electric response of root	464
172.	Diagrammatic representation of geo-electric response of root-tip	469
173.	Diagrammatic representation of geo-electric response of growing region of root	471
174.	Diagrammatic representation of the geo-perceptive layer	480
175.	The Electric Probe	483
176.	Transverse section showing continuous geo-perceptive layer (*Bryophyllum*)	488
177.	Curve of geo-electric excitation in different layers of *Nymphæa*	497
178.	Curve of geo-electric excitation in *Bryophyllum* .	497
179.	Diagram of arrangement of geotropic torsional response	503
180.	Additive effect of stimulus of gravity and light .	505
181.	Algebraical summation of geotropic and phototropic actions	505
182.	Comparative balancing effects of white and red lights	506
183.	Effect of coal gas on photo-geotropic balance .	507
184.	Diagram of magnetic balance	511
185.	Effect of variation of light on phototropic equilibrium	512
186.	Effect of variation of temperature on geotropic torsion	514
187.	Simultaneous records of variation of temperature, on up and down movement, and of torsion of the leaf of *Mimosa*	518
188.	Arrest of pulsatory movement of leaflet of *Desmodium gyrans* by light	528
189.	Effect of unilateral light on hyponastic movement .	529
190.	The Nyctitropic Recorder	537
191.	Effect of sudden darkness on leaflet of *Casia alata* .	539

FIGURE. PAGE.

192. Diurnal movement of the leaflet of *Cassia alata* . . 540
193. The day and night position of the petiole and terminal leaflet of *Desmodium gyrans* . . . 541
194. Diurnal record of the terminal leaflet of *Desmodium gyrans* 542
195. Photograph of closed flower of *Nymphæa* during day 550
196. Photograph of open flower of *Nymphæa* at night . 550
197. Action of light on the petal of *Nymphæa* . . . 551
198. Diurnal movement of the petal of *Nymphæa* . 552
199. Diurnal record of the Sijbaria Palm . . . 556
200. Diurnal record of inclined Palm, geotropically curved procumbent stem of *Tropæolum*, and diageotropic leaf of Palm 557
201. Diurnal record of leaves of *Dahlia*, *Papya*, and *Croton* 558
202. Diurnal record of procumbent stem of *Tropæolum*, and leaf of *Dahlia* for two successive days . 560
203. Abolition of the diurnal movement under constant temperature (*Tropæolum*) 565
204. Effect of inversion of plant on diurnal movement of *Tropæolum* 567
205. Electric response of the leaf stalk of *Bryophyllum* under light 571
206. Diagrammatic representation of electric after-effect of photic stimulation 571
207. After-effect of pre-maximum photic stimulation . 574
208. After-effect of maximum photic stimulation . . 574
209. After-effect of post-maximum photic stimulation . 574
210. Diurnal record of *Mimosa* in summer and winter . 577
211. Record of diurnal variation of torsion in *Mimosa* leaf 582
212. Continuous record of automatic pulsation of *Mimosa* leaf 585
213. Photometric record showing variation of intensity of light from morning to evening 586
214. Record of leaf of *Mimosa* after amputation of subpetioles 589

FIGURE. PAGE.

215. Diurnal record of *Cassia* leaf 591
216. Post-maximum after-effect of light on response of leaflet of *Cassia* 592
217. Effect of periodic alternation of light and darkness on response of *Mimosa* leaf 594
218. Pre-maximum after-effect of light in *Mimosa* . . 595
219. After-effect at maximum 595
220. Post-maximum after-effect exhibiting over-shooting below position of equilibrium 595

PART III.

TROPISM IN PLANTS.

XXII.--THE BALANCED CRESCOGRAPH

By

SIR J. C. BOSE.

WE shall in the succeeding series of papers deal with the subject of tropism in general. Different plant organs undergo curvature or bending, sometimes towards and at other times away from the stimulus which induces it. The problem is very intricate; the possibility of its solution will depend greatly on the accurate determination of the immediate and after-effects of various stimuli on the responding organ. The curvature induced in the growing organ is brought about by variation, often extremely slight, of the rate of growth; the result, moreover, is liable to be modified by the duration and point of application of stimulus. The difficulties connected with the problem can only be removed by the detection and measurement of the minutest variation in growth, and by securing a continuous and automatic record of the entire history of the change.

In the chapter on High Magnification Crescograph an account is given of the apparatus which I have devised by which the rate of growth may be magnified from ten thousand to ten millions times. It is thus possible to measure the imperceptible growth of plants for a period shorter than a single second. The variation of normal rate of growth is also found by measuring successive growth records on a stationary plate at regular intervals, say of ten seconds, or from the flexure in the growth-curve taken on a moving plate (p. 163).

I was next desirous of exalting the sensitiveness to a still higher degree by an independent method, which would not only reveal very slight variation induced in the rate of growth, but also the latent period and time-relations of the change. For this purpose I at first devised the Optical Method of Balance* which was considered at the time to be extremely sensitive. The spot of light from the Optical Lever (which magnified the rate of growth) was made to fall upon a mirror to which a compensating movement was imparted so that the light-spot after double reflection remained stationary. Any change of rate of growth—acceleration or retardation—was at once detected by the movement of the hitherto stationary spot of light in one direction or the other.

A very careful manipulation was required for the adjustment of the Optical Balance; the record moreover was not automatic. For these reasons I have been engaged for several years past in perfecting a new apparatus by which, (1) the balance could be directly obtained with the utmost exactitude, (2) where an attached scale would indicate the exact rate of growth, and (3) in which the upsetting of the balance by external stimulus would be automatically recorded, the curve giving the time relations of the change.

PRINCIPLE OF THE METHOD OF BALANCE.

I shall take a concrete example in explanation of the method of balance. Taking the rate of growth per second of a plant to be $\frac{1}{50,000}$ inch or 0·5 μ per second (equal to the wave length of sodium light), the tip of the plant will be maintained at the same point in space if we succeeded in making the plant-holder subside exactly at the same rate. The growth-elongation of the plant will then be exactly balanced by a compensating movement downwards. The state of exact balance is indicated when the

* "Plant Response" – p. 418.

recording lever of the Crescograph traces a horizontal line on the moving plate. Overbalance or underbalance will deflect the record below or above the horizontal line.

COMPENSATING MOVEMENT.

For securing exact balance the hoider of the plant P, in the given example, will have to subside at a rate of $\frac{1}{50,000}$

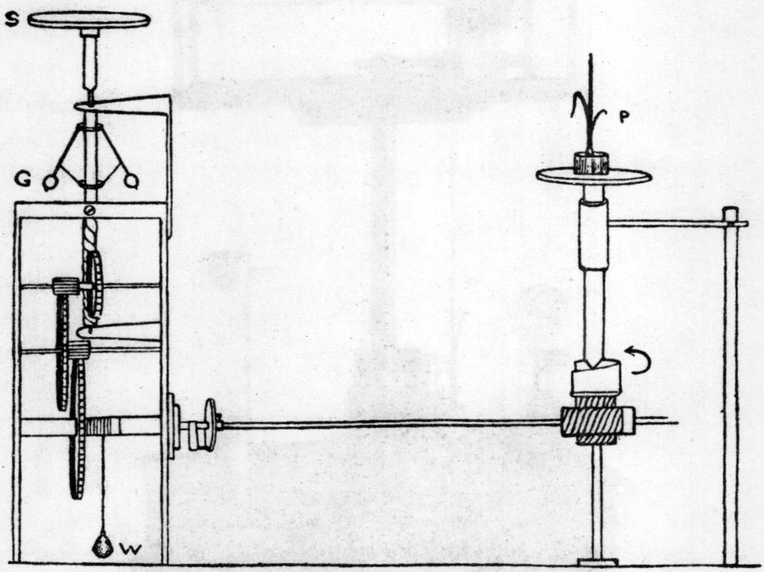

FIG. 93.—Arrangement for compensation of growth-movement by equal subsidence of plant-holder; S, adjusting screw for regulation of speed of rotation; G, governor; W, heavy weight; P, plant-holder.

inch per second. This is accomplished by a system of reducing worm and pinion, also of clock wheels (Fig. 93). The clock at first used for this purpose was worked by the usual balance wheel. Though this secured an *average* balance yet as each tick of the clock consisted of sudden movement and stoppage, it caused minute variation in the rate of subsidence; this became magnified by the Crescograph and appeared as a series of oscillations about a mean position of equilibrium. This particular defect was obviated by the substitution of a fan governor for the balance wheel. But the speed of rotation slows down with the unwinding of the main spring, and the balance obtained

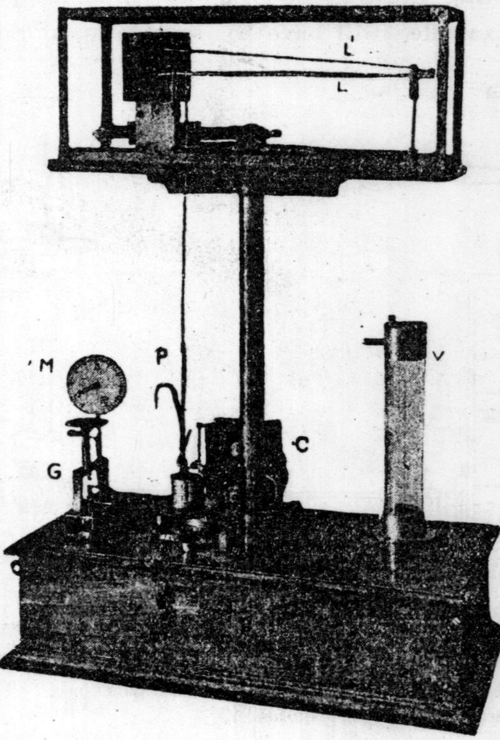

FIG. 94.—Photographic reproduction of the Balanced Crescograph. L, L[1], magnifying compound lever. R, recording plate. P, plant. C, clock work for oscillation of the plate and lateral movement. G, governor. M, circular growth-scale. V, plant-chamber.

at the beginning was found to be insufficient later on. The difficulty was finally overcome by the use of a heavy weight W, in the place of coiled spring. The complete apparatus is seen in figure 94.

For purpose of simplicity of explanation, I assumed the growth rate to have a definite value of $\frac{1}{50,000}$ inch per second. But the rate varies widely in different plants and even in the same plant at different days and seasons. In practice the rate of growth for which compensation has to be made varies from $\frac{1}{150,000}$ to $\frac{1}{25,000}$ inch, or from $0{\cdot}17\mu$ to $1{\cdot}0\mu$ per second. We have thus to secure some means of *continuous* adjustment for growth, the rate of which could be continuously varied from one to six times. This range of adjustment I have been able to secure by the compound method of frictional resistance and of centrifugal governor. As regards frictional resistance the two pointed ends of a hinged fork rub against a horizontal circular plate not shown in the figure. By means of the screw head S, the free ends of the fork spread out and the circumference of the frictional circle continuously increased. The centrifugal governor is also spread out by the action of the adjusting screw. By the joint actions of the frictional control and the centrifugal governor, the speed of rotation can be continuously adjusted from 1 to 6 times. When the adjusting screw is set in a particular position, the speed of rotation, and therefore the rate of subsidence of plant-holder, remains absolutely constant for several hours. The attainment of this constancy is a matter of fundamental importance, and it was only by the employment of the compound system of regulation that I was able to secure it.

The method of obtaining balance now becomes extremely simple. Before starting the balancing movement by clock regulation, the plant is made to record its magnified growth by the Crescograph. The compensation is effected as follows: the speed of the clockwork is at the beginning adjusted at its lowest value, and the pressure of a button starts the balancing movement of the plant downwards. On account of partial balance the record will be found to be less steep than before; the speed of the clock is gradually

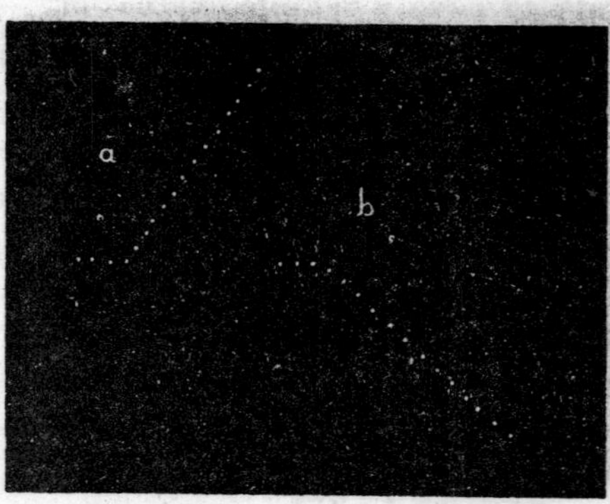

FIG. 95.—Balanced Crescographic record: (*a*) showing effect of underbalanc and (*b*) overbalance of about 3 per cent. (Magnification 2,000 times.)

increased till the record becomes perfectly horizontal under exact balance. Overbalance makes the record slope downwards. In figure 95 is seen records of underbalance (*a*) and overbalance (*b*), to the extent of about 3 per cent.

It will thus be seen that the effect of an external agent may be detected by the upsetting of the balance; an up-movement indicates (unless stated to the contrary) an enhancement of the rate of growth above the normal; and a down-movement, on the other hand, a depression of the normal rate.

Calibration.—The calibration of the instrument is obtained in two different ways. The rate of subsidence of the plant-holder, by which the balance is obtained, is strictly proportional to the rate of rotation of the vertical spindle and the attached train of clock-wheels. A striker is attached to one of the wheels, and a bell is struck at each complete revolution. The clockwork is adjusted at a medium speed, the bell striking 35 times in a minute. A microscope micrometer is focussed on a mark made on the plant-holder, and the amount of subsidence of the mark determined after one minute; this was found to be 0·0525 mm. As this fall occurred after 35 strokes of the bell the subsidence per stroke was 0·0015 mm.

Determination of the absolute rate of growth.—If growth be found balanced at N strokes of bell per minute, the rate of subsidence per second $= N \times \frac{\cdot 0015}{60}$ mm. per second

$$= N \times \cdot 000025 \text{ mm. per second}$$
$$= N \times \cdot 025\ \mu \text{ per second}$$
$$= N \times 10^{-5} \text{ inch per second.}$$

Example.—The growth of a specimen of *Zea Mays* was found balanced when the number of strokes of the bell was 20 times in a minute.

$$\text{Absolute rate of growth} = 20 \times \cdot 025\ \mu = 0{\cdot}5\ \mu \text{ per second}$$
$$\text{or} = 20 \times 10^{-5} \text{ inch} \quad ,,$$
$$\text{or} = \frac{1}{50,000} \quad ,, \quad ,,$$

If we take the wave length of sodium light λ as our standard, the growth in length per second is equal to λ. This will give us some idea of the sensitiveness of the Crescograph employed in recording the movement of growth.

GROWTH-SCALE.

The Balanced Crescograph enables us not merely to determine the absolute rate of growth, but the slightest fluctuation in that rate.

Indicator Scale.—All necessity of calculation is obviated by the scale provided with the apparatus. The speed of clockwork which brings about the balance of growth is determined by the position of the adjusting screw S, the gradual lowering of which produces a continuous diminution of speed. A particular position of the screw therefore indicates a definite rate of subsidence for balancing growth. By a simple mechanism the up or down movement of the screw causes rotation of an index pivoted at the centre of a circular scale. Each division of the scale is calibrated by counting the corresponding number of strokes of the bell per minute at different positions of the adjusting screw. The scale is calibrated in this manner to indicate different rates of growth from 0·2 μ to 1·2 μ per second.

The determination of the rate of growth now becomes extremely simple. Few turns of the screw bring about the balance of growth and the resulting position of the index against the circular scale automatically indicates the absolute rate. The procedure is even simpler and more expeditious than the determination of the weight of a substance by means of a balance.

SENSITIVENESS OF THE CRESCOGRAPHIC BALANCE.

Perhaps the most delicate method of measuring lengths is that afforded indirectly by the spectrum of a light. A good spectroscope resolves differences of wave lengths of D_1 (=0·5896 μ) and D_2 (=0·5890) *i.e.* of 1 part in a thousand. The average rate of growth of *Zea Mays* is of this order; being about 0·5 μ per second. Let us consider the question of the possibility of detecting a fractional variation of the ultra-microscopic length by means of the Balanced Crescograph. In reality the problem before us is more intricate than simple measurement of change of length; for we have to determine the *rate of variation* of length.

The sensitiveness of the balance will, it is obvious, depend on the magnifying power of the Crescograph. By the Method of Magnetic Amplification referred to in page 170, I have succeeded in obtaining a magnification of ten million times. In this method a very delicate astatic system of magnets undergoes deflection by the movement of a magnetised lever in its neighbourhood. A spot of light reflected from a small mirror attached to the astatic system, thus gives the highly magnified movement of the rate of growth, which may easily be raised to ten million times. I shall in the following describe the results obtained with this easily managed magnification of ten million times.

Determination of sensitiveness: Experiment 99.—A seedling of *Zea Mays* was placed on the Crescographic Balance; and the magnetic amplification, as stated above, was ten million times. With 18 strokes of the bell per minute the spot of light had a drift of + 266 cm. per minute to the right; this is because the growth was underbalanced. With faster rate of clock movement, *i.e.*, 21 strokes in 68 seconds or 18·53 strokes per minute, the drift of the spot of light, owing to overbalance, was to the left at the rate of − 530 cm. per minute. Thus

(1) 18 strokes per minute caused a drift of + 266 cm. per minute.

(2) 18·53 strokes per minute caused a drift of − 530 cm. per minute.

Hence by interpolation the exact balance is found to correspond to 18·177 strokes per minute.

Therefore the absolute rate of growth

$$= 18{\cdot}177 \times 0{\cdot}025\,\mu \text{ per second.}$$
$$= 0{\cdot}45\mu \text{ per second.}$$
$$= 0{\cdot}000018 \text{ inch per second.}$$

We learn further from (1) and (2) that a variation of $\frac{18{\cdot}53 - 18}{18{\cdot}177}$ produces a change of drift of the spot of light from + 266 to − 530 cm., *i.e.*, of 796 cm. per minute.

As it is easy to detect a drift of 1 cm. per minute a variation of $\frac{0 \cdot 53}{18 \cdot 177 \times 796}$, or 1 part in 27,000 may thus be detected by the Method of Balance. The spectroscopic method enabled us, as we saw, to detect change of wave length 1 part in a thousand. The sensibility of the Balanced Crescograph is thus seen to rival, if not surpass that of the spectroscope.

For obtaining a general idea of the sensitiveness, the absolute of growth in the instance given above was 0·00018 inch per second, and the Balanced Crescograph was shown capable of discriminating a variation of 1 part in 27,000; hence it is possible to detect by this means a variation of $\frac{1}{1,500}$ millionth of an inch per second.

This method of unprecedented delicacy opens out a new field of investigation on the effect of changes of environment in modification of growth; instances of this will be found in subsequent chapters. I give below accounts of certain demonstrations which will no doubt appear as very striking.

After obtaining the exact balance a match was struck in the neighbourhood of the plant. This produced a marked movement of the hitherto quiescent spot of light, thus indicating the perception of such an extremely feeble stimulus by the plant.

Breathing on the plant causes an enhancement of growth due to the joint effects of warmth and carbonic acid gas. A more striking experiment is to fill a small jar with carbonic acid and empty it over the plant. A violent movement of the spot of light to the right demonstrates the stimulating effect of this gas on growth.

The method described above is excessively sensitive; for general purposes and for the method of direct record, a less sensitive arrangement is sufficient. I give below accounts of several typical experiments in which the recording form of Crescograph was employed, the magnification being only 2,000 times.

Effect of carbonic acid on Balanced growth: Experiment 100.—I have already shown that carbonic acid diluted with air induces an enhancement of the rate of growth, but its long continued action induces a depression (p. 185). I shall now employ the Method of Balance in studying the effect of CO_2 on growth. It should be remembered in this connection that the horizontal record indicates the balance of normal rate of growth. An up-curve exhibits the induced enhancement and a down-curve, a depression of

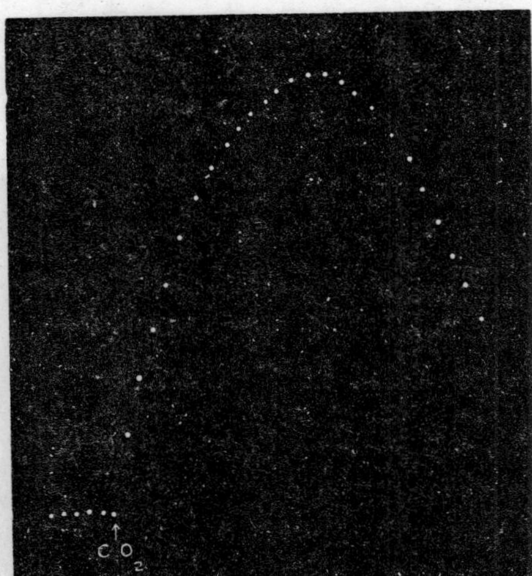

FIG. 96.—Record showing the effect of CO_2. Horizontal line at beginning indicates balanced growth. Application of CO_2 at arrow induces enhancement of growth shown by the up-curve followed by depression, shown by the down-curve. Successive dots at intervals of 10 seconds. (Seedling of wheat.)

growth. In the present experiment after obtaining the exact balance, pure carbonic acid gas was made to fill up the plant-chamber at the point marked with an arrow (Fig. 96). It will be seen that this induced an almost immediate acceleration of the rate, the latent period being less than five seconds. The acceleration continued for two and half minutes; the accelerated rate then slowed down, became enfeebled, and the growth returned for a short time to the normal as indicated by the horizontal portion at the top of the record; this proved to be the turning point of inversion from acceleration into retardation of growth. The stronger is the concentration of the gas the earlier is the point of inversion. With diluted carbonic acid the acceleration may persist for an hour or more.

EFFECT OF ANÆSTHETICS.

Effect of Ether: Experiment 101.—Dilute vapour of ether is found to induce an acceleration of rate of growth

Fig. 97.—(*a*) Effect of ether, acceleration of growth, (*b*) effect of chloroform preliminary acceleration followed by depression.

which persist for a considerable length of time. This is seen in the upsetting of the balance upwards on the introduction of the vapour (Fig. 97a.).

Effect of Chloroform: Experiment 102.—The effect of chloroform vapour is relatively more depressing than ether. Application of chloroform is seen to induce at first an acceleration which persisted for 50 seconds, but after this depression set in (Fig. 97b). Prolonged application of the anæsthetic is followed by the death of the plant.

SUMMARY.

In the Method of Balance the movement of growth upwards is compensated by an equal movement of the plant downwards, with the result that the record remains horizontal.

The effect of an external agent is immediately detected by the upsetting of the balance, up-record representing acceleration above normal, a down-record the opposite effect of depression below the normal rate.

The latent period and the after-effect of stimulus may thus be obtained with the highest accuracy.

The sensitiveness of the Method of Balance may be raised so as to indicate a variation of rate of growth smaller than $\frac{1}{1000}$ millionth of an inch per second.

XXIII.—ON TROPIC MOVEMENTS

By

SIR J. C. BOSE.

THE diverse movements induced by external stimuli in different organs of plants are extremely varied and complicated. The forces in operation are manifold—the influence of changing temperature, the stimulus of contact, of electric current, of gravity, and of light visible and invisible. They act on organs which exhibit all degrees of physiological differentiation, from the radial to the dorsiventral. An identical stimulus may sometimes induce one effect, and at other times, the precisely opposite. Thus under unilateral stimulation of light of increasing intensity, a radial organ exhibits a positive, a dia-phototropic, and finally a negative response. Strong sunlight brings about para-heliotropic or 'midday sleep' movement, by which the apices of leaves or leaflets turn towards or away from the source of illumination. The teleological argument advanced, that in this position the plant is protected from excessive transpiration, does not hold good universally; for under the same reaction, the leaflets of *Cassia montana* assume positions by which the plant risks fatal loss of water. In *Averrhoa carambola* the movement is downwards, whichever side is illuminated with strong light; in *Mimosa* leaflet the movement, under similar circumstances is precisely in the opposite direction. The photonastic movement, apparently independent of the directive action of light, has come to be regarded as a phenomenon unrelated to phototropic reaction, and due to a different kind of irritability, and a different mode of response. So very anomalous are these various effects that Pfeffer, after showing the inadequacy of different theories that have been advanced, came to the conclusion that "the precise character of the stimulatory action of light has yet to be determined. When we say that an organ curves towards a source of illumination because of its heliotropic irritability, we are simply expressing an ascertained fact in a conveniently abbreviated form, without explaining why such curvature is possible or how it is produced."*

The contradictory nature of the various responses is however not real; the apparent anomaly had lain in the fact that two definite fundamental reactions of opposite signs induced by stimulus had not hitherto been recognised and distinguished from each other. The innumerable variations in the resultant response are due to the summation of the effects of two fluctuating factors, with further complications arising from: (1) difference in the point of application of stimulus, (2) the differential excitability of the different sides of the responding organ, and (3) the effect of temperature in modifying tropic curvature. It is therefore most important to have the means for automatic record of *continuous* change in the response brought about by various factors, which act sometimes in accord, and at other times in conflict. The autograph of the plant itself, giving a history of the change in response and its time-relations, is alone decisive in explanation of various difficulties in connection with plant movements, as against the various tentative theories that have been put forward. The analysis of the resulting effect, thus rendered possible, casts new light on the phenomena of response, proving that the anomalies which had so long perplexed us, are more apparent than real.

* Pfeffer—*Ibid*—Vol. II, p. 74.

One of the causes of uncertainty lay with the question, whether response changed with the mode of stimulation. I have, however, been able to show that *all forms of stimuli* induce a definite excitatory reaction of contraction (p. 218).

Tropic movements induced by unilateral action of stimulus may, broadly speaking, be divided into two classes depending on the point of application of stimulus:

In the first, the point of application of unilateral stimulus is not on the responding organ itself, but at some distance from it. The question therefore relates to LONGITUDINAL TRANSMISSION of effect of stimulus.

In the second, unilateral stimulus acts directly on the responding organ. For the determination of the resultant movement, it is necessary to take account of effects induced on the two sides of the organ. The side adjacent to the stimulus I shall designate as the *proximal*, and the diametrically opposite as the *distal* side. The question to be investigated in this case relates to TRANSVERSE TRANSMISSION of effect of stimulus. It will be shown that the resulting movement depends on :—

(*a*) whether the tissue is a conductor or a non-conductor of excitation in a transverse direction, and

(*b*) whether it is the proximal, or the distal side of the organ that is the more excitable.

In connection with the response to environmental changes, a source of uncertainty is traceable to the absence of sufficient knowledge of the physiological effect of heat, which has been regarded as a form of stimulus: it will be shown that heat induces two distinct effects dependent on conduction and radiation. We shall in the succeeding chapters, take up the study of the physiological effects induced by changes in the environment.

XXIV.—TROPIC CURVATURE WITH LONGITUDINAL TRANSMISSION OF EFFECT OF STIMULUS

By

SIR J. C. BOSE,

Assisted by

GURUPRASANNA DAS.

I HAVE in previous chapters explained that the direct application of stimulus gives rise in different organs to contraction, diminution of turgor, fall of motile leaf, electromotive change of galvanometric negativity, and retardation of the rate of growth. I have also shown that indirect stimulation (*i.e.* application of stimulus at some distance from the responding organ) gives rise to a positive or erectile response of the responding leaf or leaflet (indicative of an increase of turgor), often followed by normal negative response. The positive impulse travels quickly. The interval of time that elapses, between the application of stimulus and the erectile response of the responding leaf, depends on the distance of the point of application, and the character of the transmitting tissue: it varies in different cases from 0·6 second to about 40 seconds. The positive is followed by a slower wave of protoplasmic excitation, which causes the excitatory fall. The velocity of this excitatory impulse is about 30 mm. per second in the petiole of *Mimosa*, and about 3 mm. per second in *Biophytum*. The positive followed by the negative thus gives rise to a diphasic response. The excitatory impulse is much enfeebled during transit: the negative impulse may thus fail to reach the responding organ, if the stimulus be feeble or if the intervening distance be long or semi-conducting. Hence moderate stimulus applied at a distance gives rise only to positive response; direct application of strong stimulus gives rise, on the other hand, to the normal negative. By employing the electric method of investigation, I have obtained with ordinary tissues the positive, the diphasic, and the negative electric response, in correspondence with the responses given by a motile organ (p. 214). The mechanics of propagation of the positive and the negative impulse are different. It is therefore necessary to distinguish the quick *transmission* of the positive impulse from the slow *conduction* of the negative impulse due to the propagation of excitatory protoplasmic change.

It should be borne in mind in this connection that all responsive movements are ultimately due to protoplasmic changes which are beyond our scrutiny. We can infer the nature of the change by the concomitant outward manifestations, which are of two kinds: the *positive*, associated with increase of turgor, expansion, and galvanometric positivity, and the *negative* with concomitant decrease of turgor, contraction, and galvanometric negativity. Thus positive and negative reactions indicate the fundamental protoplasmic changes of opposite characters.

The movement and curvature induced by stimulus have, for convenience, been distinguished as *positive curvature* (movement towards stimulus), and *negative curvature* (movement away from stimulus). Though these curvatures result from protoplasmic reactions, yet the *positive curvature* is not necessarily associated with *positive protoplasmic reaction.* It will be shown that the curvature of an organ is determined by the algebraical summation of effects induced at the proximal and distal sides of the responding organ.

Physiologists have not been aware of the dual character of the impulse generated by stimulus, and the term "transmission of stimulus" is thus misleading since its effect may be an expansion, or its very opposite, contraction. It is therefore necessary to discriminate the effect of one from the other: the impulse which induces an increase of turgor, expansion, and galvanometric positivity will be distinguished as positive, in the sense that it causes an enhancement of turgor. The other, which induces diminution of turgor and contraction, will be termed as the excitatory impulse. Transmission of the latter is dependent on conducting power of the tissue; the positive impulse is practically independent of the conducting power.

In animal physiology again, there is no essential difference between the effect of the direct and indirect stimulation. In a nerve-and-muscle preparation, for example, indirect stimulation at the nerve induces the same contraction as the direct stimulation of the muscle. The only difference lies in the latent period, which is found to be longer under indirect stimulation by the time interval necessary for the excitation to travel along the conducting nerve. It is probable that stimulus gives rise to dual impulses in the animal tissue as in the plant. But the detection of the positive impulse in the animal nerve is rendered exceedingly difficult on account of the high velocity of conduction of excitation. I have explained that the separate effects of the two impulses can only be detected if there is a sufficient lag of the excitatory negative behind the positive, so that the relatively sluggish responding organ may exhibit the two impulses one after the other. In a highly conducting tissue the lag is very slight, and the negative will therefore mask the positive by its predominant effect. In spite of the difficulty involved in the problem, I have recently been successful in demonstrating the dual impulses in the animal nerve.

In any case it is important to remember the following characteristic effects of indirect stimulation.

TABLE XXII.—SHOWING THE EFFECT OF INDIRECT STIMULATION.

Intensity of stimulus.	Character of intervening tissue	Responsive effect.
Moderate ...	Highly conducting ...	Contraction.
" ...	Non-conducting ...	Expansion.
" ...	Semi-conducting ...	Expansion followed by contraction.
Feeble ...	" " ...	Expansion.

These effects of indirect stimulation have been fully demonstrated in the case of pulvinated organs (p. 136) and growing organs (p. 215).

Having demonstrated the fundamental reactions of direct and indirect stimulation, we shall next study the tropic effects induced in growing organs by the effect of unilateral application of indirect stimulus.

Experiment 103.—I have already explained, how thermal *radiation* is almost as effective in inducing contraction and retardation of growth as the more refrangible rays of the spectrum. The thermal radiation was produced by the heating of a platinum spiral, short of incandescence, by the passage of an electric current. The intensity of radiation is easily varied by adjustment of the current by means of a rheostat. The experimental specimen was a flower bud of *Crinum.* It was held by a clamp, a little below the region of growth. Stimulus was applied below the clamp so that the transmitted effect had to pass through S, the securely held tissue (Fig. 98). A feeble stimulus was applied on one side, at the indifferent

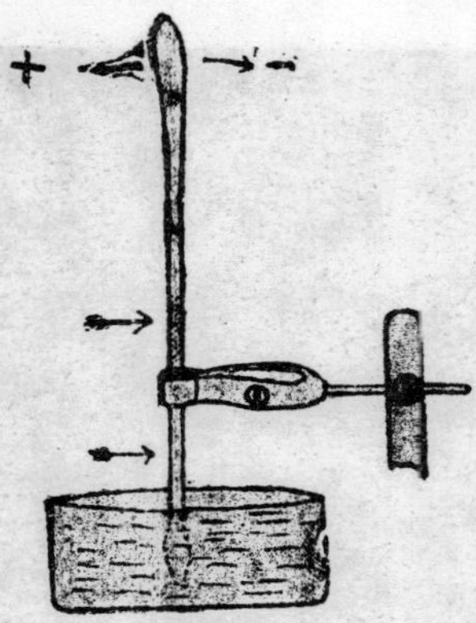

FIG. 98.—Diagrammatic representation of effects of indirect and direct stimulation. Continuous arrow represents the indirect stimulation, and the curved continuous arrow above, the induced negative curvature: dotted arrow indicates the application of direct stimulus, and the dotted curve above, the induced positive curvature.

point about 3 cm. below the region of growth. The positive effect of indirect stimulus reached the region of growth on the same side, bringing about an acceleration of growth with expansion and convexity, the resulting movement being *negative* or away from the stimulus. The latent period was ten seconds, and maximum negative movement was completed in the further course of ten seconds, after which there was a recovery in the course of 75 seconds. A stronger stimulus S′ gave a larger response; but when the intensity was raised still higher to S″, the excitatory negative impulse overtook the positive within 15 seconds of its commencement; the convex was thus succeeded by the concave curvature (Fig. 99). Direct application of stimulus at the growing region gave rise to a positive curvature.

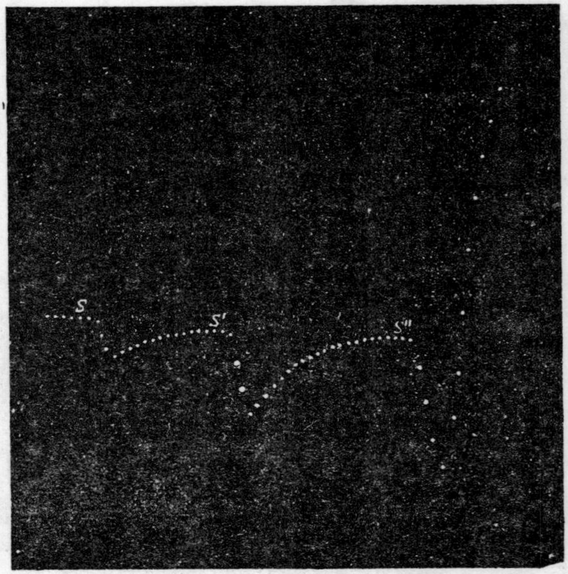

FIG. 99.—Tropic curvature of *Crinum* to unilateral indirect stimulation of increasing intensities: S, S' of moderate intensity induced negative tropic effect (movement away from the stimulated side); stronger stimulus S'' gave rise to negative followed by positive. Successive dots at intervals of 5 seconds Magnification 100 times.

The effect of feeble stimulus transmitted longitudinally is thus found always to induce convexity, *a negative curvature* and movement away from stimulus. I have obtained similar responsive movement of negative sign with various plant organs, and under various forms of stimuli. Thus in the stem of *Dregea volubilis* the longitudinally transmitted effect of light of moderate intensity was a negative curvature; direct application of light on the growing region gave, on the other hand, a positive curvature and movement towards light.

Thus while the effect of direct unilateral stimulation is a positive curvature, the effect of indirect stimulation is a negative curvature. The following table gives a summary of results of tropic effects under unilateral application of indirect stimulus.

TABLE XXIII.—SHOWING TROPIC EFFECT OF UNILATERAL APPLICATION OF INDIRECT STIMULUS.

Stimulus.	Character of intervening tissue.	Sign of tropic response.
Moderate ...	Conducting ...	Positive curvature.
...	Semi-conducting ...	Negative followed by positive.
Feeble ...	Semi-conducting ...	Negative curvature.
Moderate ...	Non-conducting ...	" "
Direct application of unilateral stimulus induces a positive curvature.		

SUMMARY.

In sensitive plants stimulus applied at a distance induces in the responding region an expansion indicative of increase of turgor.

The effect of indirect stimulation is also exhibited by an electric change of galvanometric positivity, indicative of enhancement of turgor and expansion.

Indirect stimulus induces in growing organs an enhancement of rate of growth.

Unilateral application of stimulus causes an expansion higher up on the same side to which the stimulus is applied; the result is an induced convexity, a movement away from the stimulus, *i.e.*, a negative curvature. Direct stimulus applied unilaterally at the responding region induces, on the other hand, a positive curvature.

XXV.—TROPIC CURVATURE WITH TRANSVERSE TRANSMISSION OF EFFECT OF STIMULUS

By

SIR J. C. BOSE,

Assisted by

GURUPRASANNA DAS.

WE have next to consider a very large class of phenomena arising out of the direct stimulation of one side and its transversely transmitted effect on the opposite side. The unilateral stimuli to which the plant is naturally exposed are those of contact, of light, of thermal radiation, and of gravity. There is besides the stimulation by electric current. I shall presently show that these tropic curvatures are determined by the definite effects of direct and indirect stimulations.

Under unilateral stimulus, the proximal side is found to become concave and the distal side convex; the organ thus moves towards stimulus, exhibiting a positive curvature. This movement may be due: (1) to the diminution of turgor, contraction or retardation of rate of growth of the proximal side, (2) to the increase of turgor, expansion or acceleration of rate of growth on the distal side, or (3) to the joint effects of contraction of the proximal and expansion of the distal side.

As regards the reaction of the proximal side, it has been shown that direct stimulation induces local contraction in a pulvinated organ, and retardation of growth in a growing organ. The effect induced on the distal side had hitherto remained a matter of uncertainty. In regard to this we must bear in mind that it is the effect of indirect stimulus that reaches the distal sided, inducing an enhancement of turgor and expansion of that side.

For obtaining a complete explanation of tropic curvatures in general, it is important that the induction of enhanced turgor at the distal side (by the action of stimulus at the proximal side) should be corroborated by independent methods of enquiry. One of the methods I employed for this purpose was electrical. Two electric connections were made, one with the distal point (diametrically opposite to the stimulated area), and the other, with an indifferent point at a distance. On application of stimulus of various kinds, the distal point was found to exhibit galvanometric positivity, indicative of enhancement of turgor.*

I have since been able to devise a new experiment by which the enhancement of turgor on the distal side is demonstrated in a very striking manner.

I have shown (p. 39) that the movement of the motile leaf of *Mimosa* is a reliable indicator of the state of turgor, increase of turgor inducing erection, and diminution of turgor bringing about the fall of the leaf. I shall employ the mechanical response of the leaf to demonstrate the enhancement of turgor induced by transverse transmission of effect of stimulus.

TURGOR-VARIATION UNDER TRANSVERSE TRANSMISSION OF STIMULUS-EFFECT.

Unilateral photic stimulation : Experiment 104.—A *Mimosa* plant was taken, and its stem was held vertical by means of a clamp. We apply a stimulus at a point on one side of the stem, and observe the effect of this on the state of turgor at the diametrically opposite side. In my first

* "Plant Response"—p. 519.

experiment on the subject of detection of induced change of turgor I employed the stimulus of light. A narrow beam from a small arc lamp was made to fall on the stem, at a point diametrically opposite to the motile leaf, which was to serve as a indicator for induced variation of turgor at the distal side. The leaf was attached to the recording lever, the successive dots in the record being at intervals of ten seconds. Stimulation by light caused a positive or erectile movement within 20 seconds of application. The positive response afforded a conclusive proof of the induction of an increase of turgor at the distal point.

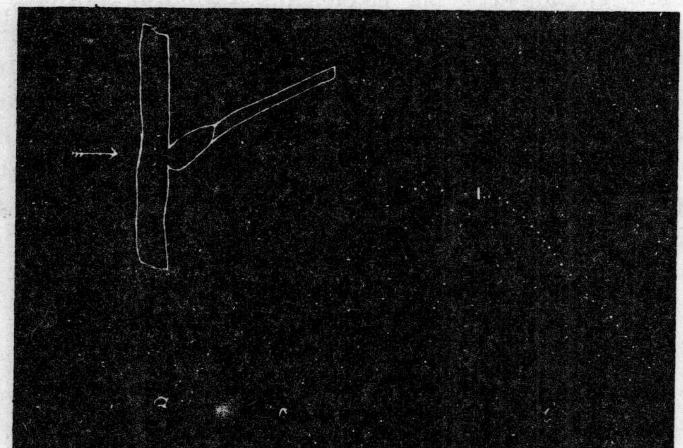

FIG. 100.—Increased turgor due to indirect stimulation, inducing erection of *Mimosa* leaf: (*a*) diagram of the experiment, point of application of stimulus indicated by arrow. (*b*) erectile response (shown by down-curve) followed by rapid fall (up-curve) due to transverse conduction of true excitation. (Successive dots at intervals of 10 seconds.)

When the stimulus is moderate or of short duration, the response remains positive. But with strong or prolonged stimulation, the slower excitatory negative impulse is conducted to the distal point and brings about the sudden fall of the leaf (Fig. 100). In the present case the excitatory impulse reached the motile organ 200 seconds after the initiation of the positive response. The stem was thin, only 2 mm. in diameter. The velocity of excitatory impulse in a transverse direction is thus 0·01 mm. per second; transverse transmission is, for obvious reasons, a much slower process than longitudinal transmission of excitation; in the *Mimosa* stem this is about 4 mm. per second.

Unilateral electric stimulation: Experiment 105.—In order to show that the effects described above are not due to any particular mode of stimulation but to stimuli in general, I carried out an additional experiment, the stimulus employed being electrical. Two fine pin-electrodes were pricked into the stem, opposite to the responding leaf of *Mimosa*; these electrodes were placed vertically one above the other, 5 mm. apart. After a suitable period, allowed for recovery from mechanical irritation, feeble tetanising electric shock was passed through the electrodes. The responsive effects at the distal side of the stem is precisely similar to those induced under unila-

FIG. 101.—Response of leaf of *Mimosa* under transverse transmission of leectric stimulus. (Compare this with fig. 100.)

teral photic stimulation; that is to say, the first effect was an erectile movement of the leaf, indicative of an induced enhancement of turgor ; the excitatory negative impulse then reached the distal point and caused a sudden fall of the leaf (Fig. 101).

The experiments that have just been described are of much significance. An organ like the stem of *Mimosa*, since it exhibits no contraction, may appear insensitive to stimulation ; but its perception of stimulus is shown by its power of transmitting two characteristic impulses, one of which is the positive, giving rise to an enhancement of turgor, and the other, the true excitatory negative, inducing the opposite reaction or diminution of turgor. Unilateral stimulation gives rise to both these effects in all organs : pulvinated, growing, and non-growing. It was the fortunate circumstance of the insertion of the motile leaf on one side of the *Mimosa* stem that enabled us to demonstrate the important facts given above.

The underlying reactions, which give rise to tropic curvature, could have been foretold from the Laws of effects of Direct and Indirect stimulation, established in previous chapters (pp. 136, 216). The resulting curvature is thus brought about by the joint effects of direct stimulation of the proximal, and indirect stimulation of the distal side. We may now recapitulate some of the important facts relating to tropic curvatures :

Indirect stimulation gives rise to dual impulses, positive and negative ; of these the positive impulse is practically independent of the conducting power of the tissue ; but the transmission of the excitatory negative impulse is dependent on the conducting power. No tissue is a perfect conductor, nor is any a perfect non-conductor of excitation, the difference is a question of degree. In a petiole or a stem the conducting power along the direction of length is considerable, but very feeble in a transverse direction. In a semi-conducting tissue, a feeble stimulus will transmit only the positive impulse ; strong or long continued stimulation will transmit both positive and negative impulses, the positive preceding the negative. The transmitted positive gives rise to increase of turgor, expansion, and acceleration of rate of growth ; the negative induces the opposite reaction of diminution of turgor, of contraction, and of retardation of rate of growth. Transverse transmission is only a particular instance of transmission in general ; the only difference is that the conducting power for *excitation* is very much less in the transverse than in the longitudinal direction. Owing to feeble transverse conductivity, the transmitted impulse to the distal side often remains positive ; it is only under strong or continued stimulation that the excitatory negative reaches the distal side and neutralises or reverses the previous positive reaction. If the distal is the more excitable side, the reversed response will appear as pronounced negative. I give a table which will clearly exhibit the effects of stimulus on the proximal and distal sides of the responding organ.

TABLE XXIV.—SHOWING RESPONSIVE EFFECTS COMMON TO PULVINI AND GROWING ORGANS UNDER UNILATERAL STIMULATION.

Effect of direct stimulation on proximal side.	Effect of indirect stimulation on distal side.
Diminution of turgor Galvanometric negativity Contraction and concavity	Increase of turgor. Galvanometric positivity. Expansion and convexity.
When stimulus is strong or long continued, the true excitatory effect is conducted to the distal side, neutralising or reversing the first response.	

The diagram which I have already given (Fig. 98) clearly explains the different tropic effects induced by changing the point of application of stimulus. We may thus have stimulus applied at the responding region itself (Direct Stimulation) or at some distance from it (Indirect Stimulation). The final effect will be modified by the conducting power of the tissue.

DIRECT UNILATERAL STIMULATION.

Type I.—The tissue has little or no power of transverse conduction : stimulus remains localised, the proximal side undergoes contraction, and the distal side expansion. The result is a positive curvature.

Type II.—The tissue is transversely conducting. Under strong and long continued stimulation the excitatory impulse reaches the distal side, neutralising or reversing the first effect.

INDIRECT UNILATERAL STIMULATION.

Type I.—The intervening tissue is an indifferent conductor : transmitted positive impulse induces expansion and convexity on the same side, thus giving rise to negative curvature (*i.e.*, away from stimulus).

Type II.—Intervening tissue is a fairly good conductor : the effect of positive impulse is over-powered by the predominant excitatory negative impulse, the final result is a concavity and positive curvature, with movement towards the stimulus.

The following is a tabular statement of the different effects induced by Direct and Indirect stimulation.

TABLE XXV.—SHOWING DIFFERENCE OF EFFECTS INDUCED BY DIRECT AND INDIRECT STIMULATION.

Stimulation.	Nature of the tissue.	Final effect.
Direct (Feeble) ...	Semi-conducting tissue.	Positive curvature.
Indirect " ...	" "	Negative curvature.
Direct (Strong) ...	Better conducting tissue.	Neutral or negative curvature.
Indirect " ...	" " "	Negative followed by positive curvature.

The results of investigations already described, enable us to formulate the general laws of tropic curvature applicable to all forms of stimuli, and to all types of responding organs, pulvinated or growing.

LAWS OF TROPIC CURVATURE.

1. (*a*) DIRECT APPLICATION OF UNILATERAL STIMULUS OF MODERATE INTENSITY, INDUCES A POSITIVE OR CONCAVE CURVATURE, BY THE CONTRACTION OF THE PROXIMAL AND EXPANSION OF THE DISTAL SIDE.

   (*b*) UNDER STRONG OR LONG-CONTINUED STIMULATION, THE POSITIVE CURVATURE IS NEUTRALISED OR REVERSED, BY TRANSVERSE CONDUCTION OF EXCITATION; THIS EFFECT IS ACCENTUATED BY THE DIFFERENTIAL EXCITABILITY OF THE TWO SIDES OF THE ORGAN.

2. (*a*) INDIRECT APPLICATION OF UNILATERAL STIMULUS OF FEEBLE INTENSITY INDUCES A NEGATIVE CURVATURE.

(*b*) IN A CONDUCTING TISSUE THE EXCITATORY EFFECT BEING TRANSMITTED UNDER STRONG AND LONG CONTINUED STIMULATION, INDUCES A POSITIVE CURVATURE.

It will thus be seen that the tropic effect is modified by :

(1) the point of application of stimulus,

(2) the intensity and duration of stimulus,

(3) the conducting power of tissue in the transverse direction,

(4) the relative excitabilities of the proximal and distal sides of the organ.

In the following series of Papers the tropic effects of various forms of stimuli will be studied in detail.

SUMMARY.

In a semi-conducting tissue Direct stimulation induces a diminution of turgor and contraction, Indirect stimulation inducing the opposite effect of increase of turgor and expansion.

Unilateral stimulation thus induces a positive curvature by the joint effects of contraction at the proximal, and expansion at the distal side.

Under strong and long continued unilateral stimulation, the excitation at the proximal side is transmitted to the distal side. Transverse conduction thus neutralises or reverses the normal positive curvature.

XXVI.—MECHANOTROPISM: TWINING OF TENDRILS

By

SIR J. C. BOSE,

Assisted by

GURUPRASANNA DAS.

IN response to the stimulus of contact a tendril twines round its support. Certain tendrils are uniformly sensitive on all sides; but in other cases, as in the tendril of *Passiflora*, the sensitiveness is greater on the under side. A curvature is induced when this side is rubbed with a splinter of wood, the stimulated under side becoming concave. This movement may be distinguished as a movement of *curling*. There is, as I shall presently show, a response where the under side becomes convex, and the curvature becomes reversed.

As regards perception of mechanical stimulus, Pfeffer discovered tactile pits in the tendrils *Cucurbitaceæ*. These pits no doubt facilitate sudden deformation of the sensitive protoplasm by frictional contact. No satisfactory explanation has however been offered as regards the physiological machinery of responsive movement. The difficulty of explanation of twining movements is accentuated by a peculiarity in the response of tendrils which is extremely puzzling. This anomaly was observed by Fitting in tendrils which are sensitive on the under side:

"If a small part of the upper side and at the same time the whole of the under side be stimulated, curvature takes place only at the places on the under side which lie opposite to the unstimulated regions of the upper side. The *sensitiveness* to contact is thus as well developed on the upper side as on the under side, and the difference between the two sides lies in the fact that while stimulation of the under side induces curvature, stimulation of the upper side induces *no visible result*, or simply inhibits curvature on the under side, according to circumstances."*

Here then we have the inexplicable phenomenon of a particular tissue, itself incapable of response, yet arresting the movement in a neighbouring tissue.

The problem before us may be thus stated: Is the movement of the tendril due to certain specific sensibility of the organ, on account of which its reactions are characteristically different from other tropic movements? Or, does the twining of tendril come under the law of tropic curvature that has been established, namely that it is brought about by the contraction of the directly stimulated proximal side, and the expansion of the indirectly stimulated distal side?

I shall now describe my investigations on the effects of direct and indirect stimulus on the growth of tendril; I

* Jost—*Ibid*—p. 490.

have in this investigation studied the effect not merely of mechanical, but also of other forms of stimuli. I shall also describe the diverse effects induced by mechanical stimulus under different conditions. From the results of these experiments I shall be able to show that the twining of the tendril comes under the general law of tropic curvature; that the curvature results from the contraction of the proximal and expansion of the distal side. Finally I shall be able to offer a satisfactory explanation of the inhibition of response of the tendril by the stimulation of the opposite side of the organ.

GENERAL EFFECTS OF INDIRECT AND DIRECT ELECTRIC STIMULATION ON THE GROWTH OF TENDRIL.

For this experiment I took a growing tendril of *Cucurbita* in which the sensitiveness is more or less uniform on all sides. The tendril was suitably mounted on the Balanced Crescograph, which records the variation of the rate of growth induced by immediate and after-effect of stimulus. The specimen is held in a clamp as in the diagram (Fig. 102), the tip being suitably attached to the recording lever. For indirect stimulation feeble shock from an induction coil is applied at the two electric connections below the clamp. Direct stimulus is applied by means of electric connections one above and the other below the clamp.

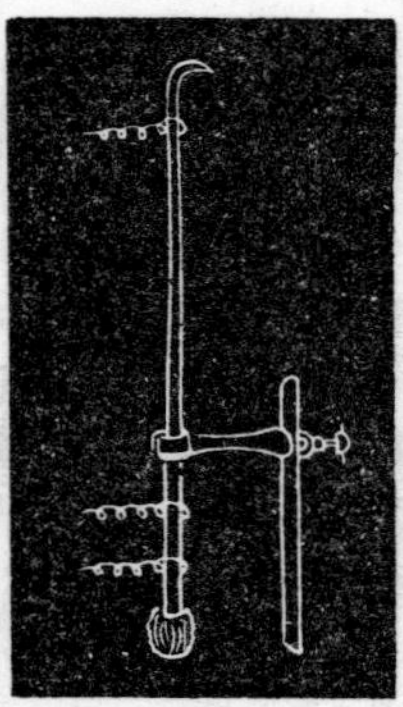

FIG. 102.—Diagrammatic representation of indirect and direct stimulation of tendril.

Effect of Indirect Stimulus: Experiment 106.—The growth of the tendril was exactly balanced, and the record became horizontal. Indirect stimulus was next applied below the clamp; this is seen to upset the balance, with the resulting up-curve which indicates a sudden acceleration of growth above the normal. This acceleration took place within ten seconds of the application of stimulus, and persisted for three minutes; after this the normal rate of growth became restored, as seen by the balanced record once more becoming horizontal (Fig. 103).

FIG. 103.—Record by Method of Balance, showing acceleration of growth of tendril (up-curve) induced by indirect stimulation. (*Cucurbita.*)

Effect of Direct Stimulus: Experiment 107.—The incipient contraction induced by direct stimulation is so great that the record obtained by the delicate method of balance cannot be kept within the plate. I, therefore, took the ordinary growth-curve on a moving plate. The first part of the curve represents normal growth; stimulus of feeble electric shock was applied at the highest point of the curve. This is seen (Fig. 104) to induce an immediate contraction and reversal of the curve which persisted for two and half minutes, after which growth was slowly renewed. The most interesting fact regarding the after-effect of stimulus is that the rate of growth became actually enhanced to three times the normal. This is clearly seen in the record (upper half of the figure) taken 20 minutes after stimulation, where the curve is far more erect than that of the normal rate of growth before stimulation.

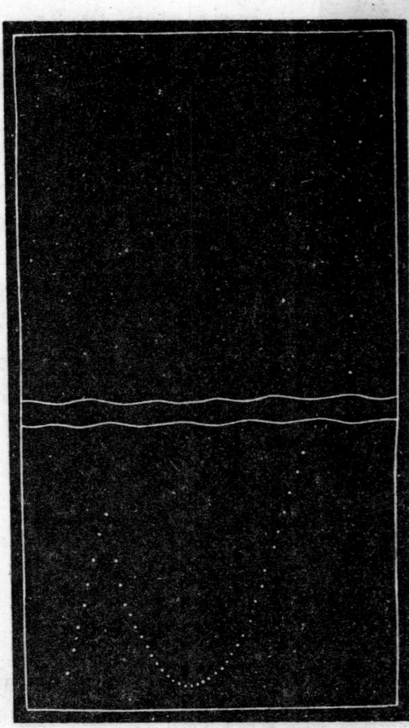

FIG. 104.—Variation of growth induced by direct stimulation. First part of the curve shows normal rate of growth. Direct stimulation induces contraction (reversal of curve). After-effect of stimulus seen in highly erect curve in upper part of record, taken 20 minutes after.

The effects of Indirect and Direct stimulation of the tendril are summarised below :

(1) Indirect stimulation induces a sudden enhancement of rate of growth, followed by a recovery of the normal rate.

(2) Direct stimulation induces a retardation of the rate of growth which may culminate into an actual contraction. *The after-effect of direct stimulus of moderate intensity is an enhancement of the rate of growth.*

The experiments described above demonstrate the effects of direct and indirect electrical stimulus. I shall now proceed to show that mechanical stimulus induces effects which are similar to those of electric stimulus.

EFFECTS OF DIRECT AND INDIRECT MECHANICAL STIMULUS.

Effect of Direct mechanical stimulus: Experiment 108.— In this case I took a tendril of *Cucurbita*, and attached it to the ordinary High Magnification Crescograph, the record of which gives the absolute rate of its normal growth, and the induced variation of that rate. The tendril was stimulated mechanically by simultaneous friction of its different sides. The immediate effect was a retardation of growth, the reduced rate being less than half the normal. There was a recovery on the cessation of the stimulus; the rate of growth was even slightly enhanced after an interval

TABLE XXVI.—SHOWING THE IMMEDIATE AND AFTER-EFFECT OF MECHANICAL STIMULATION ON TENDRIL (*Cucurbita*).

Normal rate of growth	0·44 μ per sec.
Retarded rate immediately after stimulation ...	0·20 μ " "
Recovery and enhancement after 15 minutes ...	0·50 μ " "

of 15 minutes. Table XXVI shows the immediate and after-effects of mechanical stimulation on growth.

The immediate and after-effects of mechanical stimulus on the tendril are therefore the same as that of electric stimulus. The incipient contraction under direct mechanical stimulus, moreover, is not the special characteristic of tendrils, but of growing plants in general. For I have shown (page 203) that the growth of flower stalk of *Zephyranthes* is also retarded after mechanical friction, from the normal rate 0·48 μ to 0·11 μ after stimulation. We shall find later that different plant organs, after moderate stimulation, exhibit acceleration of growth as an after-effect. The phenomenon of responsive reaction of tendril is therefore not unique, but similar to that of other organs under all forms of stimulation. The only speciality in tendril is that owing to anatomical peculiarities, the perceptive power of the organ for mechanical stimulus is highly developed.

We are now in a position to offer an explanation of the induced concavity of the stimulated side of the tendril, and its recovery after brief contact. The experiments that have been described show that :

(1) the proximal side contracts because it is directly stimulated, and the distal side, being indirectly stimulated, expands; the curvature is thus due to the joint effects of contraction of one side, and expansion of the opposite side, and

(2) the recovery of the tendril after brief contact is hastened by the after-effect of stimulus, which is expansion and acceleration of growth.

The results given above will also be found to explain Fitting's important observations* that (*a*) the stimulated side of the tendril undergoes transient contraction with subsequent acceleration of growth, and that (*b*) the distal or convex side undergoes an immediate enhancement of growth.

* Pfeffer—*Ibid*—Vol. III, p. 57.

I give below a record given by a tendril of *Cucurbita* in response to unilateral contact of short duration (Fig. 105). Successive dots in the record are at intervals of three seconds. The latent period was ten seconds, and the maxi-

FIG. 105.—Positive curvature of tendril of *Cucurbita* under unilateral stimulus of contact at ×.

mum curvature was attained in the course of two and a half minutes. The curvature persisted for a further period of two minutes after which recovery was completed in the course of 12 minutes. Feeble stimulation is attended by a recovery within a short period, but under strong stimulus the induced curvature becomes more persistent.

INHIBITORY ACTION OF STIMULUS.

I have referred to the remarkable observation of Fitting that though the application of stimulus on the upper side of the tendril of *Passiflora* did not induce any response, yet it inhibited the normal response of the under side.

The results of experiments which I have described will, however, afford a satisfactory explanation of this curious inhibition. It has been explained, that the curvature of the tendril is due to the joint effects of diminished turgor and contraction at the directly stimulated side, and an enhancement of turgor and expansion on the opposite side. In the diagram seen in figure 106, the left is the more excitable side, and contraction will induce concavity of the stimulated side. But if the opposite or less excitable side of the tendril be stimulated at the same time, then the transmitted effect of indirect stimulus will induce enhancement of turgor and expansion on the left side, and thus neutralise the previous effect of direct stimulus. An inhibition of response will thus result from the stimulation of the opposite side.

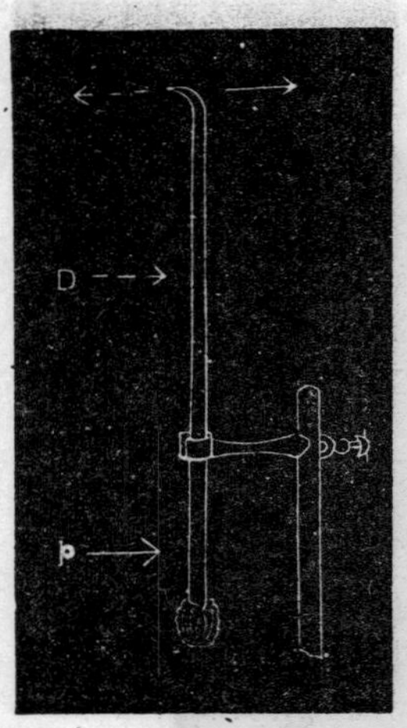

FIG. 106.—Diagrammatic representation of effects of Indirect and Direct unilateral stimulation of the tendril. Indirect stimulation, I, induces movement away from stimulated side (negative curvature) represented by continuous arrow. Direct stimulation, D, induces movement towards stimulus (positive curvature) indicated by dotted arrow.

A difficulty arises here from the fact that the upper side of the tendril (the right side in Fig. 106) is supposed to be inexcitable and non-contractile. Hence there may be a misgiving that the stimulation of the non-motile side may not induce the effect of indirect stimulus (an increase of turgor and expansion) at the opposite side, which is to inhibit the response. But I have shown that even a non-contractile organ under stimulus generates both the impulses, positive and negative. This is seen illustrated in figure 100, where the rigid stem of *Mimosa* was subjected to unilateral stimulation; the effect of indirect stimulus was found to induce an enhancement of turgor at the diametrically opposite side, and thus caused an erectile movement of the motile leaf. Electric investigations which I have carried out also corroborate the results given above. Here also stimulation of a non-motile organ at any point, induces at a diametrically opposite point, a positive electric variation indicative of enhanced turgor. It will thus be seen that inhibition is possible even in the absence of contraction of the upper side of the tendril; hence the contraction of the directly stimulated side is neutralised by the effect of indirect stimulation of the distal side.

RESPONSE OF LESS EXCITABLE SIDE OF THE TENDRIL.

It is generally supposed that the upper side of the tendril of *Passiflora* is devoid of contractility. This is however not the case, for my experiments show that stimulation of the upper side also induces contraction and concavity of that side, though the actual movement is relatively feeble.

Experiment 109.—In order to subject the question to quantitative test I applied feeble stimulus of the same intensity on upper and lower side alternately. Successive stimuli were kept more or less uniform by employing the following device. I took a flat strip of wood 1 cm. in breadth, and coated 2 cm. of its length with shellac varnish mixed with fine emery powder. On drying the surface became rough, the flat surface was gently pressed against the area of the tendril to be stimulated, and quickly drawn so that the rough surface 2 cm. × 1 cm. was rubbed against the tendril in each experiment. Stimulation, thus produced, induced a respon e movement of each side of the organ. The extent of the maximum movement was measured by the microscope micrometer. The following results were obtained with four different specimens.

TABLE XXVII.—SHOWING THE RELATIVE INTENSITIES OF RESPONSES OF THE UPPER AND UNDER SIDE OF TENDRIL (*Passiflora*).

Movement induced by stimulation of under side, A.	Movement induced by stimulation of upper side, B.	Ratio $\frac{B}{A}$.
(1) 85 divisions ...	14 divisions ...	1/6
(2) 106 " ...	15 " ...	1/7
(3) 60 " ...	8 " ...	1/7
(4) 80 " ...	10 " ...	1/8

It will thus be seen that the upper side of the tendril is not totally inexcitable, its power of contraction being about one-seventh that of the under side.

NEGATIVE CURVATURE OF THE TENDRIL.

I shall now describe certain remarkable results which show that under certain definite conditions the tendril moves away from the stimulated side. I have explained, how in growing organs the effect of unilateral stimulus longitudinally transmitted, induces an expansion higher up on the same side to which the stimulus is applied, resulting in convexity and movement away from the stimulus (cf. Laws of Tropic Curvatures, p. 286). As the reaction of tendril is in no way different from that of growing organs in general, it occurred to me that it would be possible to induce in it a negative curvature by application of indirect unilateral stimulus.

Experiment 110.—A tendril of *Passiflora* was held in a clamp, as in the diagram (Fig. 106) in which the left is the more excitable side of the organ. The responsive movement of the tendril is observed by focussing a reading microscope on a mark on the upper part of the tendril. Direct mechanical stimulation at the dotted arrow makes the tendril move in the same direction, the response being *positive.* But if stimulus be applied on the same side below the clamp the tendril is found to move away from stimulus, the response being now *negative.* This reversal of response, as previously stated, is due to the fact that the transmitted effect of indirect stimulus induces an acceleration of growth higher up on the same side, which now becomes convex. The result though unexpected, is in every way parallel to the response of the flower bud of *Crinum*, in which the normal positive response was converted into negative by changing the point of application of stimulus, so that it became indirect (p. 216).

SUMMARY.

The response of tendril is in no way different from that of growing organs in general.

Direct stimulus, electrical or mechanical, induces an incipient contraction ; the after-effect of a feeble stimulus is an acceleration of growth above the normal. Indirect stimulus induces an enhancement of the rate of growth.

Under unilateral mechanical stimulus of short duration the directly excited proximal side undergoes contraction, the indirectly stimulated distal side exhibits the opposite effect of expansion. The induced curvature is thus due to the joint effects of the contraction of one side, and the expansion of the opposite side.

As the after-effect of direct stimulus is an acceleration of growth above the normal, the stimulated side undergoes an expansion by which the recovery is hastened.

Unilateral application of direct stimulus induces a *positive* curvature, but the same stimulus applied at a distance from the responding region induces a *negative* curvature.

The tendril of *Passiflora* is excitable both on the upper and under sides; the excitability of the under side is about seven times greater than that of the upper side.

Stimulation of one side of the tendril induces an expansion of the opposite side, even in cases where the contractility of the stimulated side is feeble.

The response to stimulation of the more excitable side of the tendril is thus inhibited by the stimulation of the opposite side. This is because of the neutralisation of the effect of direct by that of indirect stimulation.

XXVII.—ON GALVANOTROPISM

By

SIR J. C. BOSE,

Assisted by

GURUPRASANNA DAS.

BEFORE describing the effect of unilateral application of an electrical current in inducing tropic curvature, I shall give an account of the polar effect of anode and cathode on the pulvinated and growing organs. In my previous work* on the action of electrical current on sensitive pulvini I have shown that :—

(1) at the 'make' of a current of moderate intensity a contraction takes place at the cathode; the anode induces no such contractile effect;

(2) at the 'make' of a stronger current both the anode and cathode induce contraction.

I have also carried out further investigations on the polar effect of current on the autonomous activity of the leaflet of *Desmodium gyrans.* These rhythmic pulsations can be recorded by my Oscillating Recorder. Each pulsation consists of a sudden contractile movement downwards, corresponding to the systole of a beating heart, and a slower up movement of diastolic expansion. Application of

* "Irritability of Plants." p. 212.

cathode at the pulvinule was found to exert a *contractile* reaction, exhibited either by the reduction of normal limit of diastolic expansion, or by an arrest of movement at systole. The effect of anode was precisely the opposite; the induced *expansion* was exhibited either by reduction of normal limit of systolic contraction, or by arrest of pulsation at diastole.

From the above results it is seen that with a feeble current:

(1) contraction is induced at the cathode, and

(2) expansion is brought about at the anode.

These effects take place under the action of a feeble current. Under strong currents, contraction takes place both at the anode and the cathode.

POLAR EFFECT OF ELECTRICAL CURRENT ON GROWTH.

The object of this investigation was to determine whether anode and cathode exerted similar discriminative and opposite effects on growth. For this experiment I took a specimen of *Kysoor* and determined the region where growth was maximum. A piece of moist cloth was wrapped round this region to serve as one of the two electrodes. The second electrode was placed in the neighbouring indifferent region where there had been a cessation of growth.

Effect of Cathode: Experiment 111.—The particular specimen of *Kysoor* had a normal rate of growth of 0·48 μ per second. On application of the cathode the rate was reduced to 0·14 μ per second, or to less than a third. This will be seen in record (Fig. 107), where N is the normal rate of growth and K, retarded rate under the action of the cathode.

Fig. 107.

Fig. 108.

FIG. 107.—Retardation of rate of growth under the action of cathode (*Kysoor*).
FIG. 108.—Acceleration of rate of growth under anode (*Kysoor*).

Effect of anode : Experiment 112.—If the cathode induced a retardation, the anode might be expected to induce an acceleration of growth. But in my first experiment on the action of anode, I could detect no perceptible variation of rate of growth. In trying to account for this failure, I found that the specimen employed for the experiment had normally a very rapid rate of growth. It appeared that an induced acceleration would be brought out more conspicuously by choosing a specimen in which the growth-rate was low, rather than in one in which it was near its maximum. Acting on this idea, I took another specimen of *Kysoor* in which the normal rate was as slow as 0·10 μ per second. On applying the anode to the growing region, there was an enhancement to one and half times the normal rate (Fig. 108).

TABLE XXVIII.—EFFECT OF ANODE AND CATHODE ON GROWTH (*Kysoor*).

Specimen A	Normal rate ... ... Acceleration under anode ...	0·10 μ per sec. 0·155 μ per sec.
Specimen B	Normal rate ... ... Retardation under cathode ...	0 48 μ per sec. 0·14 μ per sec.

The effects given above take place under the action of a feeble current. Strong current on the other hand induces a retardation or an arrest of growth.

I have in the above experiments demonstrated the normal effect of anode in inducing expansion and acceleration of rate of growth ; the cathode was shown to induce contraction and retardation of growth. Unilateral application of anode and cathode thus induces appropriate curvatures in pulvinated and in growing organs.

SUMMARY.

The effects of an electric current on growth is modified by the direction of current. A feeble anodic current enhances the rate of growth ; a cathodic current on the other hand induces a retardation of the rate. Strong current, both anodic and cathodic, induces a retardation.

XXVIII.—ON THERMONASTIC PHENOMENA

By

SIR J. C. BOSE,

Assisted by

SURENDRA CHANDRA DAS.

IN describing thermonastic curvatures Pfeffer says that "a special power of thermonastic response has been developed by various flowers, in which low temperatures produce closing movements, and high temperatures, opening ones. The flowers of *Crocus vernus* and *Crocus luteus* are specially responsive, as also those of *Tulipa Gesneriana* for these flowers perceptibly respond to a change of tepmerature of half a degree centigrade."*

We have hitherto studied the response of various organs to *stimulus*; we have now to deal with the effect of thermal variation. Does rise of temperature act like other forms of stimuli or is its action different? We have therefore to find:

(1) The physiological effect of variation of temperature.

(2) Whether thermonastic irritability is confined only to certain classes of organs, or is it a phenomenon of very wide occurrence?

(3) Whether variation of temperature induces in anisotropic organs only one type of response, or two types, positive and negative.

* Pfeffer—*Ibid*, Vol. III, p. 112.

(4) The law which determines the direction of responsive movement.

EFFECT OF RISE OF TEMPERATURE.

As regards the effect of rise of temperature we have seen that, within normal limits, it induces expansion and acceleration of the rate of growth. Stimulus, on the other hand, induces precisely the opposite effect. Hence the physiological reaction of steady rise of temperature is, generally speaking, antagonistic to that of stimulus. This conclusion is supported by numerous experiments which I have carried out with various plant organs. Example of this will be found in the present and subsequent chapters.

DIFFERENT THERMONASTIC ORGANS.

The only condition requisite for the exhibition of response is the differential excitability of an anisotropic organ. It is therefore likely to be exhibited by a large variety of plant organs, such as pulvini, petioles, leaves, and flowers, and my results show that this is actually the case. This particular sensibility, moreover, is not confined to delicate structures, but is extended to rigid trees and their branches.

Before proceeding further, it is necessary to draw attention to the confusion which arises from the use of the common prefix '*thermo*' in thermonasty and thermo-tropism. With regard to this Pfeffer says "It is not known whether radiated and conducted heat exercise a similar thermotropic reaction."* I shall show that the reactions to radiant heat, and to conducted heat

* Pfeffer—*Ibid*, Vol. III, p. 177.

(rise of temperature) are of opposite character, radiation inducing contraction, and rise of temperature, expansion. It is therefore advisable to distinguish the thermal, or temperature effect, from the radio-thermal effect of infra-red radiation.

TWO TYPES OF RESPONSE.

As regards the effect of variation of temperature I shall proceed to show that there are two distinct types, which I shall, for convenience, distinguish as the *Positive* and *Negative.*

Positive thermonastic reaction is exhibited by organs in which the upper half is the more excitable. Response to rise of temperature is by *downward* or *outward* movement. In floral organs this finds expression by a movement of opening. In illustration of this may be cited the examples of the well known *Crocus* and also of *Zephyranthes.*

Negative thermonastic movement is shown by organs in which the lower half is the more excitable. Here the response to rise of temperature is by an *upward* or *inward* movement. I shall show that an example of this is furnished by the flower of *Nymphœa* which closes under rise, and opens during fall of temperature.

POSITIVE THERMONASTIC RESPONSE.

Response of Zephyranthes : *Experiment 113.*—Viewed from the top, the inner side of the petal of a flower is the upper side. The *Crocus* flower under rise of temperature opens outwards by expansion of the inner side, which must be the more excitable. As *Crocus* was not available in Calcutta, I found the flower of *Zephyranthes* (sometimes called the Indian *Crocus*) reacting to variation of temperature in a manner similar to that of *Crocus*, that is to say, the flower opens under rise and closes with a fall of temperature. For obtaining record all the perianth segments but one was removed. This segment was attached to the recording lever. On lowering of temperature through 5°C. there was an up-movement, or a movement of closure.

FIG. 109.--Thermonastic and radionastic responses of petal of *Zephyranthes* C, closing movement due to cooling, and H, opening movement due to warming; R, closing movement due to heat-radiation. Note opposite responses to rise of temperature and to thermal radiation.

Rise of temperature induced, on the other hand, a movement of opening.

Effect of thermal radiation : Experiment 114.—I stated that the effect of thermal radiation acts as a stimulus, inducing a reaction which is antagonistic to that of rise of temperature. In verification of this, I subjected the specimen to the action of infra-red radiation acting from all sides. The result is seen in the responsive movement of closure (Fig. 109 R). These experiments demonstrate clearly that the responses to rise of temperature and thermal radiation are of opposite signs.

As a movement of closure was induced by the diffuse stimulus of thermal radiation, it is evident that this must have been brought about by the greater contraction of the inner half of the perianth ; hence the inner half of the organ is relatively the more excitable.

NEGATIVE THERMONASTIC RESPONSE.

Response of Nymphæa : *Experiment 115.*—Many of the Indian *Nymphœaceœ* have their sepals and petals closed

FIG. 110.—The Thermonastic Recorder. T, metallic thermometer attached to the short arm of the upper lever ; the specimen of *Nymphæa*, N, has one of its perianth leaves attached to the short arm of the second lever by a thread. C, clock work for oscillation of the plate.

during the day, and open at night. I find that the perianth leaves of this flower are markedly sensitive to variation of temperature. The Thermonastic Recorder employed in this investigation is shown in figure 110. The record given in figure 111 shows that the perianth segment, subjected to a few degrees' rise of temperature, responded by an up-movement of closure, due to greater expansion of the outer half. The latent period was 6 seconds, and the maximum effect was attained in the further course of 21 seconds. This experiment shows that the thermonastic response of this flower is of the negative type.

Effect of stimulus: Experiment 116.—In the positive type of thermonastic organs, where rise of temperature induced a movement of opening, stimulus induced the opposite movement of closure (Expt. 114). We shall now study the effect of stimulus on the movement of *Nymphæa*, which undergoes closure during rise of temperature, as seen

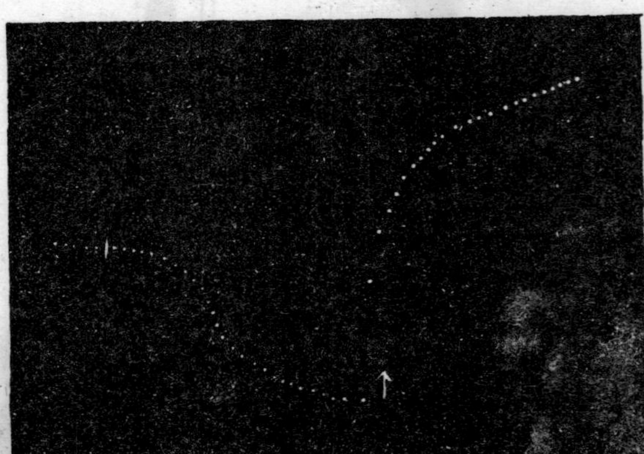

FIG. 111.—Negative thermonastic response of *Nymphæa*. Application of warmth at the vertical mark induced up-movement of closure, but stimulus of electric shock at arrow induced rapid excitatory down movement of opening. Successive dots at intervals of a second.

in the first part of the record in figure 111. Stimulus of electric shock was applied at the point marked with an arrow; the response is seen to be by a movement of opening. Here also we find the effects of rise of temperature and of stimulus to be antagonistic to each other. This will be clearly seen in the following tabular statement.

TABLE XXIX.—SHOWING THE EFFECT OF RISE OF TEMPERATURE AND OF STIMULUS ON THERMONASTIC ORGANS.

Specimen.	Effect of rise of temperature.	Effect of stimulus.
Zephyranthes (positive type).	Movement of opening ...	Movement of closure.
Nymphæa (negative type).	Movement of closure ...	Movement of opening.

In *Nymphæa* it is the outer side of the perianth that is relatively the more excitable since diffuse electric stimulus induces a movement of opening due to the greater contraction of the outer side. It is by the greater expansion of this more excitable side that the movement of closure is effected during rise of temperature.

From the results of experiments given above we arrive at the following :—

LAW OF THERMONASTIC REACTION

RISE OF TEMPERATURE INDUCES A GREATER EXPANSION OF THE MORE EXCITABLE HALF OF AN ANISOTROPIC ORGAN.

SUMMARY.

Thermonastic movements are induced by the differential physiological effect of variation of temperature on the two halves of an anisotropic organ.

Rise of temperature induces greater expansion, and enhancement of rate of growth of the more excitable half of the organ; lowering of temperature induces the opposite effect.

Two types of thermonastic movements are met with, the *positive* exhibiting a movement of opening during rise of temperature; in these the inner half of the organ is relatively the more excitable. Example of this is seen in the *Crocus* and in *Zephyranthes*.

In the *negative* type, rise of temperature induces a movement of closure. Here the outer half of the organ is the more excitable. The flower of *Nymphæa* belongs to this type.

The effect of stimulus is antagonistic to that of rise of temperature. In positive thermonastic organs stimulus induces a movement of closure; in the negative type it induces a movement of opening.

XXIX.—ON PHOTOTROPISM

By

SIR J. C. BOSE.

IN different organs of plants the stimulus of light induces movements of an extremely varied character. Radial organs exhibit tropic movements in which the position of equilibrium is definitely related to the direction of incident stimulus. Nastic movements under the action of light are, on the other hand, regarded as curvatures of the organ which show "no relation to the stimulus but is determined by the activity of the plant itself".* There are thus two classes of response to light which seem to be unrelated to each other. Returning to the directive action of light, radial stems often bend towards the light, while certain roots bend away from it. It may be thought that this difference is due to specific difference of irritability between shoot and root, the irritability of the former being of a positive, and of the latter, of a negative character. But there are numerous exceptions to this generalisation. Certain roots bend towards the light, while a stem, under different circumstances, moves towards light or away from it. Again an identical organ may exhibit a positive or a negative curvature. Thus the leaflets of *Mimosa pudica* acted on by light from above fold upwards, the phototropic effect being *positive.* But the same leaflets acted on by light from below exhibit a folding upwards, the phototropic effect being now *negative.* Effects precisely the opposite are found with the leaflets of *Biophytum* and *Averrhoa.* They fold downwards whether light acts from

* Jost—*Ibid,* p. 428.

above or below. Finally, a radial organ is found to exhibit under light of increasing intensity or duration, a positive, a dia-phototropic, or a negative phototropic curvature.

In these circumstances the theory of specific positive and negative irritabilities is untenable; in any case, it throws no light on the phenomenon of movement. The difficulties of the problem are thus clearly stated by Pfeffer: "When we say that an organ curves towards a source of illumination, because of its heliotropic irritability and we are simply expressing an ascertained fact in a conveniently abbreviated form, without explaining why such curvature is possible or how it is produced . . . . Many observers have unfortunately devoted their attention to artificially classifying the phenomenon observed, and have entirely neglected the explanation of causes underlying them."*

COMPLEXITY OF PROBLEM OF PHOTOTROPIC REACTION.

The complexity of phototropic reaction arises from the summated effects of numerous factors; for explanation of the resultant response it is therefore necessary to take full account of the individual effect of each of them.

Among these operative factors in phototropic reaction may be mentioned :—

(1) The difference of effects induced by light at the proximal and distal sides of the organ.

(2) The modification of the latent period with the intensity of stimulus.

(3) The after-effect of stimulus.

(4) The modifying influence of tonic condition on response.

(5) The effect of direction of light.

* Pfeffer—*Ibid*, Vol. II, p. 74

(6) The effect of intensity of light.

(7) The effect of duration of stimulation.

(8) The transmitted effect of light.

(9) The effect of unequal excitability in different zones of the organ.

(10) The effect of transverse conduction in modification of the sign of response.

(11) The effect of temperature on phototropic action.

(12) The modification of response due to differential excitability of the organ.

(13) Nastic and tropic reactions.

(14) The torsional effect of light.

The sketch given above will give us some idea of the complexity of the problem. In this and in the following papers I shall describe the investigations I have carried out on the subjects detailed above.

ACTION OF LIGHT.

I have shown that there is no essential difference between the responses of pulvinated and growing organs, that diminution of turgor induced by stimulus brings about contraction in the one, and retardation of the rate of growth in the other. Indirect stimulation, on the other hand, induces an expansion and acceleration of the rate of growth. The experimental investigation on the tropic effect of light may therefore be carried out both with pulvinated and growing organs.

As regards the effect of direct stimulus of light on growing organs we found (p. 208) that it induces an incipient contraction, seen in diminution of the rate of growth; this incipient contraction culminates in an actual contraction under increasing intensity of light. The contraction under direct stimulation is also observed in pulvinated organs. When light acts from above the upper half of the pulvinus undergoes contraction, resulting in erection of the motile leaf or leaflets. As regards the effect of indirect unilateral stimulus of light on the distal side of the organ, we found that its effect is an enhancement of turgor (p. 281). Hence the positive tropic curvature under light is brought about, as in the case of other forms of stimuli, by the contraction of the proximal, and expansion of the distal sides of the organ.

Various analogies have been noticed between phototropic and geotropic reactions, and it has been supposed that the two phenomena are closely related to each other. This has even led to assumption that there are phototropic particles which function like statoliths in geotropic organs. There is, however, certain outstanding difference between the two classes of phenomena. In the case of light, the incident energy is entirely derived from the outside. But in geotropism, the force of gravity by itself is ineffective without the intervention of the weight of cell-contents to exert pressure on the sensitive ectoplasm, and thus induce stimulation. This aspect of the subject will be treated in greater detail in a subsequent chapter.

POSITIVE PHOTOTROPIC CURVATURE.

I shall now describe the phototropic effect of unilateral light in pulvinated, and in growing organs. From the explanation that has already been given, it will be understood that the side of the organ directly acted on by light undergoes contraction and concavity.

Tropic curvature of pulvinated organs: Experiment 117.—For this experiment I employed the terminal leaflet of the bean plant. The source of illumination was 32 c.p. electric lamp, enclosed in a metallic tube with circular

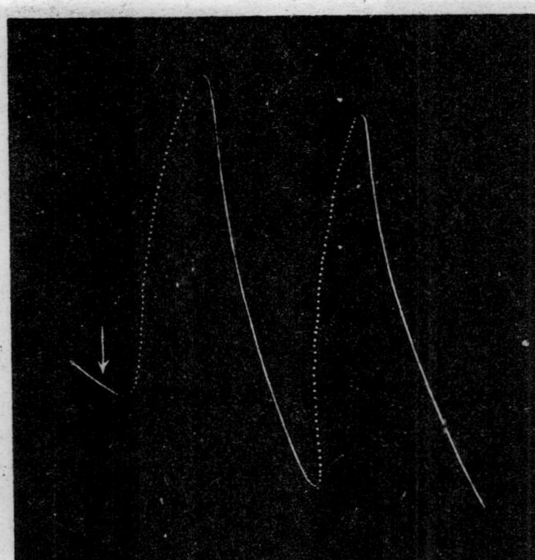

FIG. 112.—Successive positive responses of the terminal leaflet of bean. Light applied from above for 20 seconds ; complete recovery in 8 minutes.

aperture for passage of light. The leaflet was attached to an Oscillating Recorder. Light was applied on the upper half of the pulvinus for 20 seconds ; this induced an up-movement of the leaflet, due to the contraction of the upper half of the organ. Recovery took place in course of 8 minutes (Fig. 112).

POSITIVE PHOTOTROPIC CURVATURE OF GROWING ORGANS.

Effect of moderate stimulation : Experiment 118.—I shall presently show that the intensity of phototropic reaction depends on the intensity and duration of the incident light. A moderate and effective stimulation may thus be produced by short exposure to strong light. For my

FIG. 113.

FIG. 114.

FIG. 113.—Positive curvature under moderate phototropic stimulation. Note complete recovery (*Dregea*).

FIG. 114.—Persistent positive curvature under stronger stimulation (*Dregea*).

present experiment I took a stem of *Dregea volubilis*, and applied light from a small arc lamp to one side of the organ for 1 minute ; this induced a positive curvature followed by complete recovery on the cessation of light (Fig. 113).

Effect of strong stimulation : Experiment 119.—After recovery of the stem of the last experiment, the same light was applied for 5 minutes. It is seen that the curvature is greatly increased (Fig. 114). Thus the phototropic curvature increases, within limits, with the duration of stimulation. The curvature induced under stronger stimulation remained more or less persistent. In certain instances there was a partial recovery after a considerable length of time; in others curvature was fixed by growth.

PHENOMENON OF RECOVERY.

On the cessation of stimulus of moderate intensity the heliotropically curved organ straightens itself; similar effects are also found in other tropic curvatures. Thus a tendril straightens itself after curvature induced by contact of short duration. The theory of rectipitality has been proposed to account for the recovery, which assumes the action of an unknown regulating power by which the organ is brought back to a straight line; but beyond the assumption of an unknown specific power, the theory affords no explanation of the mechanism by which this is brought about.

The problem before us is to find out the means by which the organ straightens itself after brief stimulation. It will also be necessary to find out why there is no recovery after prolonged stimulation. We have thus to investigate the after-effect of stimulus of various intensities on growth, and the Balanced Method of recording Growth offers us an unique opportunity of studying the characteristic after-effects.

IMMEDIATE AND AFTER-EFFECT OF LIGHT ON GROWTH.

As regards the effect of light I have already shown

(1) that a sub-minimal stimulus induces an acceleration of growth, but under long continued action the acceleration is converted into normal retardation (p. 225),

(2) that a stimulus of moderate intensity induces the normal retardation of the rate of growth.

It is evident that there is a *critical intensity* of stimulus, above which there is a retardation, and below which there is the opposite reaction of acceleration. This critical intensity, I have found to be low in vigorous specimens, and high in sub-tonic specimens. Thus the same intensity of stimulus may induce a retardation of growth in specimens the tonic condition of which is *above par*, and an acceleration in others, in which it is *below par*. The following experiments will demonstrate the immediate and after-effect of light of increasing intensity and duration.

Effect of light of moderate intensity: Experiment 120.—The source of light was a small arc lamp placed at a distance of 50 cm., the intensity of incident light was increased or decreased by bringing the source of light nearer

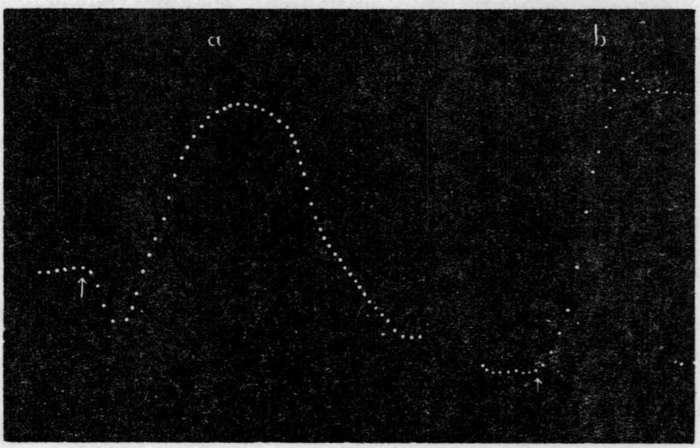

FIG. 115.—Immediate and after-effect of stimulus of light on growth. (*a*) shows immediate effect of moderate light to be a transitory acceleration (down-curve) followed by retardation (up-curve). The after-effect on cessation of light is an acceleration (down-curve) followed by restoration to normal. (*b*) Immediate and after-effect of stronger light: immediate effect, a retardation; after-effect, recovery to normal rate without acceleration.

or further away from the plant. Two inclined mirrors were placed behind the plant so that the specimen was acted on by light from all sides. A seedling of wheat was mounted on the Balanced Crescograph, and record was first taken under exact balance; this gives a horizontal record. The up-curve represents retardation, and down-curve acceleration of rate of growth. The source of light was at first placed at a distance of 50 cm. from the plant, and exposure was given for 4 minutes at the point marked with an arrow (Fig. 115a). We shall find in the next chapter that the *intensity of phototropic effect is proportional to the quantity of incident light.* This quantity at the beginning proved to be subminimal, and hence there was an acceleration at the beginning. Continued action induced the normal effect of retardation, as seen in the subsequent resulting up-curve. On the cessation of light, the balance was upset in an opposite direction, the resulting down-curve showing an acceleration of the rate of growth above the normal. This acceleration persisted for a time, after which the normal rate of growth was restored, as seen in the curve becoming once more horizontal. *The after-effect of light of moderate intensity is thus a temporary acceleration of rate of growth above the normal.*

Effect of strong light: Experiment 121.—The same specimen was used as in the last experiment. By bringing the source of light to a distance of 25 cm. the intensity of light was increased fourfold; the duration of exposure was kept the same as before. The record (Fig. 115b) shows that a retardation of rate of growth occurred from the very beginning without the preliminary acceleration. This is for two reasons: (1) the increased intensity was now above the critical minimum, and (2) the tone of the organ had become improved by previous stimulation. On the cessation of light, the after-effect showed no enhancement of rate of growth, the recovery from retardation to the normal rate being gradual. In the next experiment (the result of which is not given in the record) the intensity of light was increased still further; the retardation now became very marked, and it persisted for a long time even on the cessation of light.

We thus find that:

(1) The immediate effect of light of moderate intensity is a preliminary acceleration, followed by normal retardation. The acceleration is the effect of sub-minimal stimulation. The immediate after-effect is an acceleration above the normal.

(2) The immediate effect of strong light is a retardation from the beginning; the immediate after-effect shows no acceleration, the growth rate being gradually restored to the normal.

(3) Under very strong light the induced retardation is very great, and this persists for a long time even on the removal of light.

The experiments described explains the reasons of complete recovery after moderate stimulation, and also the absence of recovery after strong stimulation. The immediate after-effect of moderate stimulation is shown to be an acceleration of rate above the normal. Returning to tropic curvature, the contraction at the proximal side induced by unilateral light is thus compensated by the accelerated rate of growth on the cessation of light. There is no such compensation in the case of strong and long continued action of light; for the after-effect of strong light shows no such acceleration as the immediate after-effect.

We may perhaps go a step further in explaining this difference. Stimulus was found to induce at the same time two physico-chemical reactions of opposite signs (p. 144).

One is the 'up' or A-change, associated with increase of potential energy of the system, and the other is associated with 'down' or D-change, by which there is a run-down or depletion of energy. With moderate stimulation the A-and-D effects are more or less comparable to each other. But under strong stimulation the down-change is relatively greater. Hence on cessation of moderate stimulation the increase of potential energy, associated with A-change, finds expression in enhancement of the rate of growth. The depletion of energy under strong stimulation is, however, too great to be compensated by the A-change.

LATENT PERIOD OF PHOTOTROPIC REACTION.

With reference to the latent period Jost thus summarizes the known results:* "The latent period of the heliotropic stimulus has already been determined. According to Czapek it amounts to 7 minutes in the cotyledons of *Avena* and in *Phycomyces;* 10 minutes in hypocotyls of *Sinapis alba* and *Beta vulgaris*, 20 minutes in the hypocotyl of *Helianthus*, and 50 minutes in the epicotyl of *Phaseolus*. If one of these organs be unilaterally illuminated for the specified time, heliotropic curvature ensues afterwards in the dark, that is to say, we meet with an after-effect in this case as in geotropism. We are quite ignorant, however, as to whether and how the latent period is dependent on the intensity of light."

With regard to the question of relation of the latent period to the intensity of stimulus I have shown (p. 166) that the latent period is shortened under increasing intensity of stimulus. In the case of tropic curvature induced by light, I find that the latent period is reduced under

* Jost—*Ibid*, p. 473.

increasing intensity of light. The shortest latent period found by Czapek, as stated before, was 7 minutes. But by employing high magnification for record, I find that the latent period of phototropic action under strong light to be a question of seconds.

Determination of the latent period: Experiment 122.—I give a record of response (Fig. 116) of the terminal leaflet of *Erythrina inidca* to light acting from above. The recording plate was made to move at a fast rate, the

FIG. 116.—Latent period for photic stimulation at vertical line. Successive dots at intervals of 2 seconds. (*Erythrina indica*).

successive dots being at intervals of 2 seconds. The latent period in this case is seen to be 35 seconds. By the employment of stronger light I have obtained latent period which is very much shorter.

The term latent period is used in two different sense. It may mean the interval between the application of stimulus and the initiation of response. In the experiment described above, the latent period is to be understood in this sense. But in the extract given above, Jost uses the term latent period as the shortest period of exposure necessary to induce phototropic reaction as an after-effect. What then is the shortest exposure that will induce a retardation of growth? For this investigation I employed the very sensitive method of the Balanced Crescograph.

GROWTH-VARIATION BY FLASH OF LIGHT FROM A SINGLE SPARK.

Experiment 123.—I stated that the more intense is the light, the shorter is the latent period. The duration of a single spark discharge from a Leyden jar is almost instantaneous, the duration of discharge being of the order of $\frac{1}{100,000}$th of a second. The single discharge was made to take place between two small steel spheres, the light given out by the spark being rich in effective ultra-violet rays. The plant used for the experiment was a seedling of wheat. It was mounted on the Balanced Crescograph, and its normal growth was exactly compensated as seen

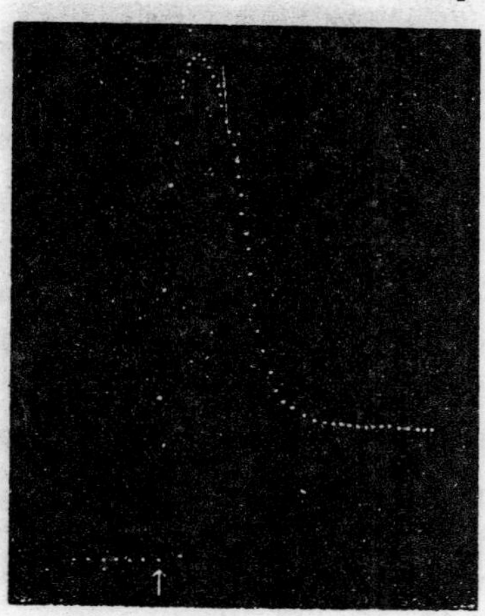

FIG. 117.—Effect of a single electric spark on variation of growth. Record taken by Balanced Crescograph. Up-curve shows induced retardation of growth; the after-effect is an acceleration (down-curve) followed by restoration to normal.

in the first part of the record. The spark gap was placed at a distance of 10 cm. from the plant; there was the usual arrangement of inclined mirrors for illumination of the plant. The flash of light from a single spark is seen to induce a sudden retardation of rate of growth which lasted for one and half minutes. The record (Fig. 117) shows another interesting peculiarity of acceleration as an after-effect of moderate stimulation. After the retardation which lasted for 90 seconds, there is an acceleration of growth above the normal, which persisted for 6 minutes, after which the rate of growth returned to the normal.

In order to show that the induced variation is due to the action of light and not to any other disturbance, I interposed a sheet of ebonite between the spark-gap and the plant. The production of spark produced no effect, but the removal of the ebonite screen was at once followed by the characteristic response.

MAXIMUM POSITIVE CURVATURE UNDER CONTINUED ACTION OF LIGHT.

The positive curvature is, as we have seen, due to the contraction of the proximal side and expansion of the distal side. The curvature will increase with growing contraction of the proximal side; a maximum curvature is however reached since:

(1) the contraction of the cells must have a limit,

(2) the bending organ offers increasing resistance to curvature, and

(3) the induced curvature tends to place the organ parallel to the direction of light when the tropic effect is reduced to a minimum.

The pulvinus of *Erythrina* exemplifies the type of reaction in which the positive curvature reaches a maximum, (see below Fig. 132) beyond which there is no further change. This is due to absence of transverse conductivity in the organ. The modifying effect of transverse conductivity on response will be dealt with in the next chapter.

SUMMARY.

The positive phototropic curvature is brought about by the joint effects of the directly stimulated proximal, and indirectly stimulated distal side.

The phototropically curved organ undergoes recovery after brief stimulation.

The recovery after moderate stimulation is hastened by the previously stimulated side exhibiting an acceleration of the rate of growth above the normal. The after-effects of photic and mechanical stimulation are similar.

The latent period of photic reaction is shortened with the increasing intensity of light. The seedling of wheat responds to a flash of light from an electric spark, the duration of which is about a hundred thousandth part of a second.

Tissues in which the power of transverse conduction is negligible, the positive phototropic curvature under continued action of light attains a maximum without subsequent neutralisation or reversal.

XXX.—DIA-PHOTOTROPISM AND NEGATIVE PHOTOTROPISM

By

SIR J. C. BOSE,

Assisted by

GURUPRASANNA DAS.

I have explained how under the action of unilateral light the positive curvature attains a maximum. There are, however, cases where under the continued action of strong light the tropic movement undergoes a reversal. Thus to quote Jost: "Each organism may be found in one of the three different conditions determined by the light intensity, *viz.* (1) a condition of positive heliotropism, (2) a condition of indifference, (3) a condition of negative heliotropism"*. No explanation has however been offered as to why the same organ should exhibit at different times, a positive, a neutral, and a negative irritability. These changing effects exhibited by an identical organ is thus incompatible with the theory of specific sensibility, assumed in explanation of characteristic differences in phototropic response.

In regard to this I would draw attention to an important factor which modifies the tropic response, namely, the effect of transverse conduction of excitation. I shall presently describe in detail a typical experiment of the

* Jost—*Ibid*—p. 462.

effect of unilateral stimulus of light on the responsive movement of main pulvinus of *Mimosa pudica*. The results will be found of much theoretical interest, since a single experiment will give an insight to all possible types of phototropic response. Before describing the experiment I shall demonstrate the tropic reactions of the two halves of the pulvinus of *Mimosa*.

UNEQUAL EXCITABILITY OF UPPER AND LOWER HALVES OF PULVINUS TO PHOTIC STIMULATION.

I have by method of selective amputation shown that as regards electric stimulation the excitability of the upper half of the pulvinus is very much less than that of the lower half (p. 85). I have obtained similar results with photic stimulation.

Tropic effect of light acting from above: Experiment 124.—Light of moderate intensity from an incandescent electric lamp was applied on the upper half of the pulvinus of *Mimosa* for 4 minutes; this induced a contraction of the stimulated upper half and gave rise to an up or erectile response. On the stoppage of light recovery took place in the course of ten minutes. The phototropic curvature is thus seen to be positive. A series of such positive responses of the upper half of the pulvinus is given in figure 118.

Effect of light acting from below: Experiment 125.—Light was now applied from below; this also induced a contraction of the lower half of the pulvinus, causing a down-movement (Fig. 119). As the responsive movement is towards light, the phototropic effect must be regarded as positive. The greater excitability of the lower half of the pulvinus is shown by the fact that the response of the lower half of the pulvinus to ten seconds' exposure is

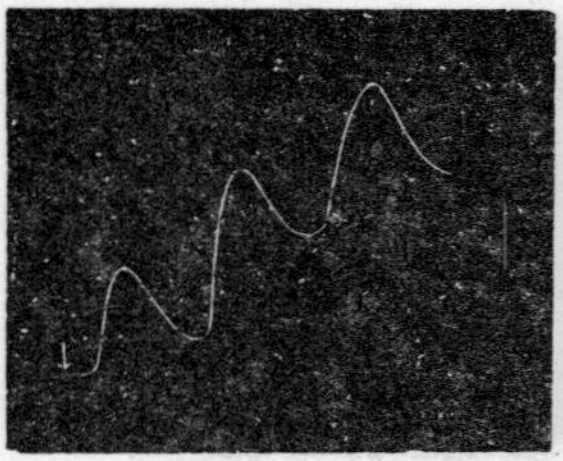

FIG. 118.

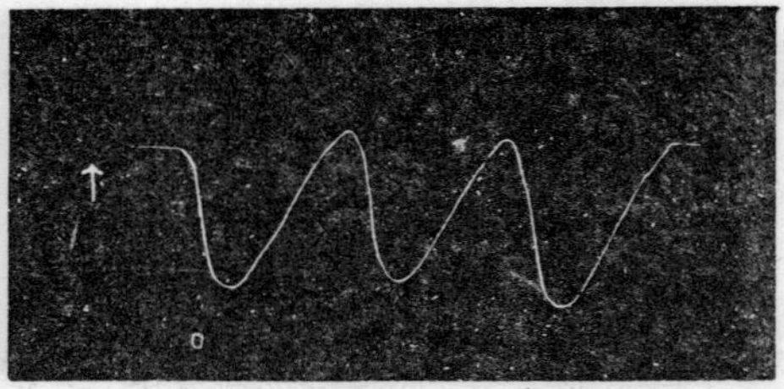

FIG. 119.

FIG. 118.—Series of up-responses of *Mimosa* leaf to light applied on upper half of pulvinus.

FIG. 119.—Down-responses given by the same plant on application of light from below.

even larger than that given by the upper half under the prolonged exposure of 240 seconds.

TRANSFORMATION OF POSITIVE TO NEGATIVE PHOTOTROPIC CURVATURE.

Experiment 126.—A beam of light from a small arc lamp was thrown on the upper half of the pulvinus. After a latent period of 5 seconds, a positive curvature was initiated, by the contraction of the upper and expansion of the lower side of the organ. But under continued action of light, the excitatory impulse reached the lower half of the organ, causing a rapid fall of the leaf, and a *negative* curvature. The arrival of transmitted excitation at the more excitable distal half of the organ is clearly demonstrated by the very rapid down-movement, seen as the up-curve in the record (Fig. 120). In sensitive specimens this movement is so abrupt and rapid, that the writing lever is jerked off above the recording plate before making a dot on it. The thickness of the pulvinus was 1·5 mm., the distance which the excitatory impulse has to traverse to reach the lower half would thus be about 0·75 mm. The period for transverse transmission

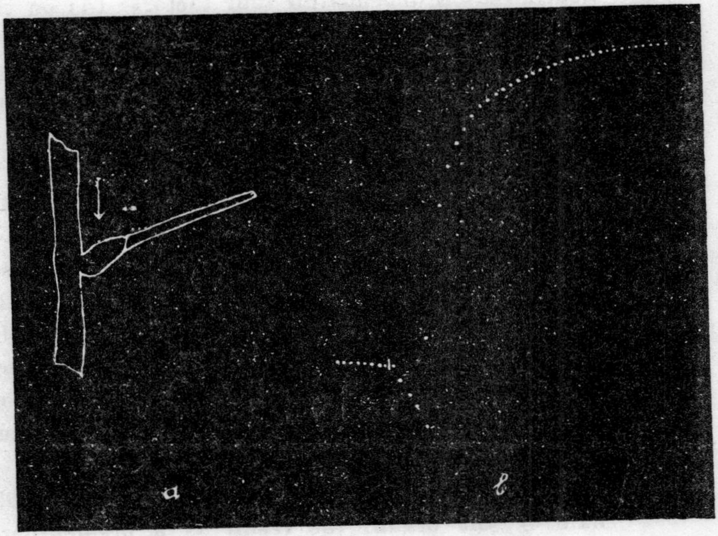

FIG. 120.—Record of effect of continuous application of light on upper half of pulvinus of *Mimosa* leaf. Note erectile response (positive curvature) followed by neutralisation and pronounced reversal into negative due to transverse conduction of excitation. Up-movement shown by down curve, and *vice versâ*.

of excitation under strong light was found to vary in different cases from 50 to 80 seconds. The velocity of transmission of excitation in a transverse direction through the pulvinus is about 0·011 mm. per second, which is not very different from 0·010 mm. per second in the stem (p. 282).

Returning to the main experiment we find that :

(1) As a result of unilateral action of light, there was positive phototropic curvature which lasted for 50 seconds.

(2) Owing to the internal conduction of excitation the positive effect underwent neutralisation by the excitatory contraction of the distal side. This neutralisation depends on four factors: (*a*) on the intensity of the stimulus, (*b*) on the conductivity of the organ in a transverse direction, (*c*) on the thickness of the intervening tissue, and (*d*) on the relative excitability of the distal as compared to the proximal side. The extent of positive curvature also depends on the pliability of the organ.

(3) In anisotropic organs where the distal side is physiologically the more excitable than the proximal, the internally diffused excitation brings about a greater contraction of the distal, and the *positive* phototropic curvature becomes reversed to a very pronounced *negative*. The effect of the internally diffused stimulus is thus the same as that of external diffuse stimulus.

(4) When the stimulus is applied on the more excitable half of the organ, the result is a predominant contraction of that half, which cannot be neutralised by the excitation conducted to the less excitable half of the organ. As the curvature is towards the stimulus, the phototropic curvature thus remains positive, even under continued stimulation.

The positive curvature is due to the differential action of unilateral stimulus on the proximal and distal sides. But when a strong light is made to act continuously on one side of an organ, the excitation becomes internally diffused, and the differential effect on the two sides is reduced in amount or vanishes altogether. Owing to the weak transverse conductivity of the tissue, while the effect of a feeble stimulus remains localised, that of a stronger stimulus is conducted across it.

Oltmanns found that the seedling of *Lepidium sativum* assumed a transverse or dia-phototropic position under intense and long continued action of light of 600,000 Hefner lamps. He regards this as the indifferent position. But the neutralisation of curvature is not, as explained before, due to a condition of indifference, but to the antagonistic effects of the two opposite sides of the organ, the proximal being stimulated by the direct, and the distal by the transversely conducted excitation. I obtained such neutralisation with *Dregea volubilis* under the prolonged unilateral action of arc-light. The first effect was positive; this was gradually and continuously neutralised under exposure for two hours; even then the neutralisation was not complete. I shall presently adduce instances where the neutralisation was not merely complete, but the final effect was an actual reversal into negative response.

SUPPOSED PHOTOTROPIC INEFFECTIVENESS OF SUNLIGHT.

I may here consider the remarkable fact that has been observed, but for which no explanation has been forthcoming, that "direct sunlight is too bright to bring about heliotropic curvature, only diffuse, not direct sunlight has the power of inducing heliotropic movements."* But we cannot conceive of light suddenly losing its phototropic effect by an increase of intensity. The experiment just described will offer full explanation for this apparent anomaly. Feeble or moderate stimulus remains, as we have seen, localised, hence the contraction of the proximal side gives rise to positive curvature. But the intense excitation caused by sunlight would be transmitted to the distal side and thus bring about neutralisation. It is the observation of the final result that has misled observers as to the inefficiency of direct sunlight. A continuous record of the response of the organ shows, on

* Jost—*Ibid*—p. 464.

the other hand, that the first effect of strong light is a positive curvature, and that under its continuous action the positive effect becomes neutralised (cf. Fig. 121). In the study of phototropic action, the employment of strong light has many advantages, since the period of experiment is, by this means, materially shortened. The continuous record then gives an epitome of the various phases of reaction.

NEGATIVE PHOTOTROPISM.

I shall next show the continuity of responsive phototropic effects, from the positive curvature to the negative, through the intermediate phase of neutralisation. I have in the preceding paragraph described an experiment where under a given intensity and duration of exposure the excitations of the proximal and distal sides bring about neutralisation, the organ assuming a dia-phototropic position. If the intensity or duration of the stimulating light be further increased, it is easy to see that while excitation transmitted to the distal side is being increased, the excitatory contraction on the proximal side may, at the same time, be decreased owing to fatigue brought on by over-stimulation.

In connection with this it should be borne in mind that the pulvinus of *Mimosa* exhibits under continuous stimulation, a fatigue relaxation instead of normal contraction. Similar effects are known to take place in animal muscles. The effect of relatively greater excitation will thus give rise to negative phototropic curvature. The transverse conductivity of organs of diverse plants will necessarily be different. The neutralisation and reversal into negative will thus depend on three factors: the transverse conductivity of the organ, the intensity, and duration of stimulus.

Neutralisation and reversal under increased intensity of light: Experiment 127.—It is advisable to employ thin specimens (in which the transverse distance is small) for the exhibition of reversal effect. I took a hypocotyl of *Sinapis nigra* and subjected it to unilateral action of light from a 16 candle-power incandescent electric lamp placed at a distance of 10 cm. A maximum positive curvature was induced in the course of 50 minutes. The intensity of light was afterwards increased by bringing the lamp nearer to a distance of 6 cm. This resulted in a process of neutralisation of the preceding response; after an exposure of 70 minutes the specimen assumed a dia-phototropic position in which it remained in equilibrium. Sunlight was next applied, and in the further course of 30 minutes there was a pronounced reversal into negative phototropic curvature.

Neutralisation and reversal under continuous stimulation: Experiment 128.—In the last experiment the different changes in the response were brought about by successive increase in the intensity of light. In the present experiment, very strong light was applied from the beginning, and continuous record was taken of the change in the response. In order to reduce the period of experiment I employed a mercury vapour lamp which emits the most effective violet and ultra-violet rays. The specimen used was a seedling of the rice plant (*Oryza sativa*). The first effect of light was a positive curvature which attained its maximum; after this there was a neutralisation in less than six minutes after the application of light. The further

FIG. 121.—Positive and negative phototropic responses of *Oryza* under continued unilateral stimulus of intense light from arc lamp.

continuation of light induced a pronounced negative curvature (Fig. 121).

I shall in the next chapter give other instances which will show that all organs (pulvinated and growing) possessed of power of transverse conduction, exhibit a transformation of response from positive to negative under continued action of strong light.

Thus an identical organ, under different conditions of intensity and duration of stimulus, exhibits *positive* phototropic, *dia*-phototropic, and *negative* phototropic curvatures, proving conclusively that the three effects are not due to three distinct irritabilities. The responsive movements are, on the other hand, traced to a fundamental excitatory reaction, remaining either localised or increasingly transmitted to the distal side.

NEGATIVE PHOTOTROPISM OF ROOTS.

From the analogy of opposite responses of shoot and root to stimulus of gravity, it was surmised that the root would respond to light by a negative curvature. This was apparently confirmed by the negative phototropic curvature of the root of *Sinapis*. The supposed analogy is however false; for while the stimulus of gravity acts, in the case of root, only on a restricted area of the tip, the stimulus of light is not necessarily restricted in the area of its action. That there is no true analogy between the action of light and gravitation is seen from the fact that while gravitation induces in the root a movement opposite to that in the stem, in the case of light, this is not always so; for though a few roots turn away from light, others move towards the light.

As regards negative phototropic response of the root of *Sinapis*, it will be shown (p. 376) to be brought about by algebraical summation of the effects of direct and indirect photic stimulus.

SUMMARY.

The normal positive phototropic curvature is modified by transverse conduction of true excitation to the distal side of the organ.

The extent of neutralisation or reversal due to internal conduction of excitation from the proximal to the distal side of the organ depends: (*a*) on the intensity of the incident stimulus, (*b*) on the conductivity of the organ in a transverse direction, (*c*) on the thickness of the intervening tissue, and (*d*) on the relative excitability of the distal as compared to the proximal side.

The dia-phototropic position is not one of indifference, but of balanced antagonistic reactions of two opposite sides of the organ.

The supposition that direct sunlight is phototropically ineffective is unfounded. The response is fully vigorous, but the first positive curvature may in certain cases be neutralised by the transmission of excitation to the distal side.

Under light of strong intensity and long duration, the transmitted excitation to the distal side neutralises, and finally reverses the positive into negative curvature.

The *positive*-phototropic, the *dia*-phototropic, and the *negative* phototropic curvatures are not due to three distinct irritabilities but are brought about by a fundamental excitatory reaction remaining localised or increasingly transmitted to the distal side.

XXXI.—THE RELATION BETWEEN THE QUANTITY OF LIGHT AND THE INDUCED PHOTOTROPIC CURVATURE

By

SIR J. C. BOSE,

Assisted by

SURENDRA CHANDRA DAS, M.A.

I SHALL in this chapter describe experiments in support of the important proposition that *the intensity of phototropic action is dependent on the quantity of incident light.* The proportionality of the tropic effect to the quantity of light will be found to hold good for the median range of stimulation; the deviation from this proportionality at the two ends of the range of stimulation—the sub-minimal and supra-maximal—is, as we shall find, capable of explanation, and will be fully dealt with in the next chapter.

The quantity of light incident on the responding organ depends: (1) on the intensity of light, (2) on the angle of inclination or *the directive angle*,* and (3) on the duration

* The directive angle θ is the angle of inclination of the rays of light to the responding surface. The angle θ is complementary to the angle of incidence i in optics. $\sin\theta = \cos i$.

of exposure. I shall give a detailed account of the investigation relating to the individual effects of each of these factors on the tropic reactions not merely in pulvinated but also in growing organs.

EFFECT OF INCREASING INTENSITY OF LIGHT ON TROPIC CURVATURE.

The intensity of light was increased in successive experiments, in arithmetical progression 1 : 2 : 3 by suitably diminishing the distance between the plant and the source of light, and the resulting tropic curvatures recorded.

Effect of increasing intensity of light on the pulvinus of Desmodium gyrans : *Experiment 129.*—The source of light was a 50 candle-power incandescent lamp, and the duration of exposure was 1 minute. The specimen employed was a terminal leaflet of *Desmodium gyrans* (Fig. 122) the pulvinus of which is very sensitive to light. It is more convenient to manipulate a cut specimen of the leaf, instead of the whole plant. The petiole is placed in water contained in a U-tube; the depressing effect of wound passes off in the course of an hour or so. Light of increasing intensity is applied from above; this induces a contraction of the upper half of the pulvinus, and

FIG. 122.—Leaf of *Desmodium gyrans*, with the terminal large, and two lateral small leaflets. These latter exhibit automatic pulsations.

the resulting response is recorded by means of the Oscillating Recorder (Fig. 123).

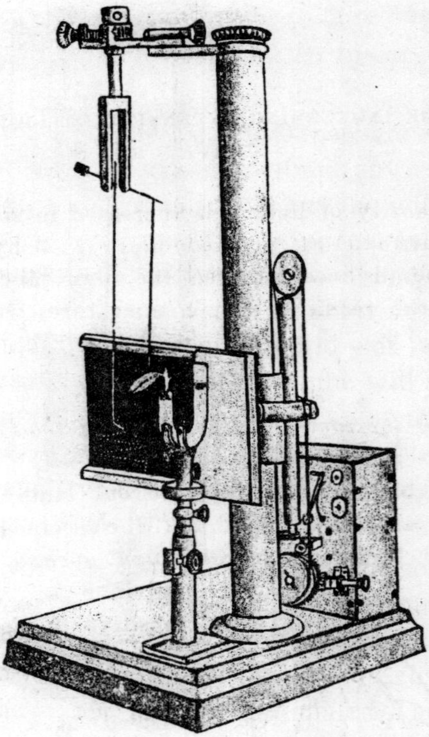

FIG. 123.—The Oscillating Recorder (From a photograph).

The first record was obtained under a given intensity, and the second, under an intensity twice as great. The tropic effects are seen to increase with the intensity

(Fig. 124). If the tropic curvature increased proportionately to the intensity, the two responses should have been

FIG. 124.

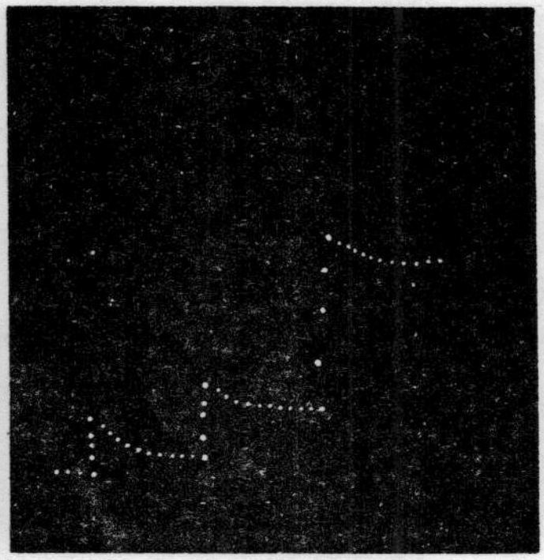

FIG. 125.

FIG. 124.—Tropic effect of increasing intensity of light 1 : 2 ; on the response of terminal leaflet of *Desmodium gyrans.*

FIG. 125.—Tropic effect of increasing intensity of light 1 : 2 : 3 on growing organ (*Crinum*).

in the ratio of 1 : 2 ; the actual ratio was however slightly greater, *viz.* 1 : 2·6. In this connection it will be shown in the next chapter, that strict proportionality holds good only in the median range, and that the susceptibility for excitation undergoes an increase at the beginning of the phototropic curve.

Effect of increasing intensity of light on the tropic curvature of growing organs.—As the tropic curvature is primarily due to the retardation of growth induced by light at the proximal side of the organ, it will be of interest to recapitulate the results I obtained (p. 208) on the effects of increasing intensity of light on growth itself. The normal rate of growth of the specimen in the dark was 0·47 μ per second ; this was reduced to 0·29 μ under an intensity of one unit, to 0·17 μ under two, and to 0·10 μ under three units. Growth became arrested when the intensity was raised to four units. Thus increasing intensity of light induces an increasing retardation of growth at the proximal side of the organ. This aided by the effect of indirect stimulus at the distal side brings about an increasing positive curvature.

Experiment 130.—The flower bud of *Crinum* was used for the experiment, the source of light being a small arc lamp. The duration of exposure was one minute. Increasing intensity of light gave rise to increasing positive curvatures (Fig. 125) in the ratio of 1 : 2·5 : 5 under increasing intensities which varied as 1 : 2 : 3.

THE EFFECT OF INCREASING ANGLE.

The quantity of light which falls on an unit area of the responding organ varies as sin θ where θ is the directive angle *i. e.* the angle made by the rays with the surface. Some allowance has to be made for the amount of light reflected from the surface, this being greater at 45° than at 90°.

Tropic response of pulvinus of Desmodium gyrans : *Experiment 131.*—For application of light at various angles an incandescent electric lamp was mounted at one end of a brass tube, a collimating lens being placed at the

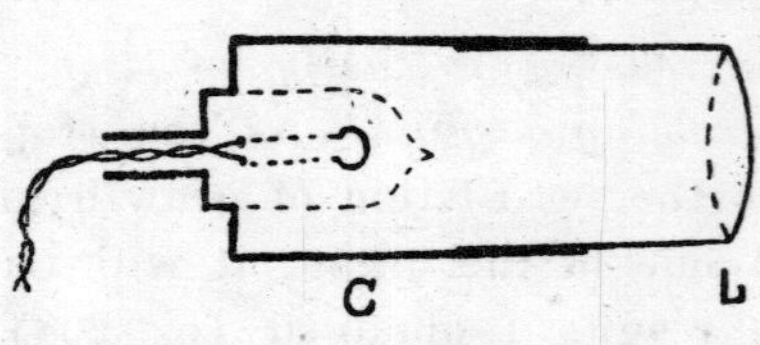

FIG. 126.—The Collimator.

other (Fig. 126). The parallel beam of light from the collimator could be sent at various angles by rotating the collimator tube round an axis at right angles to the tube.

The specimen employed was the terminal leaflet of *Desmodium gyrans*; light was applied for a minute in the two successive experiments for the two angles of 45° and 90°. The record (Fig. 127) shows that the phototropic

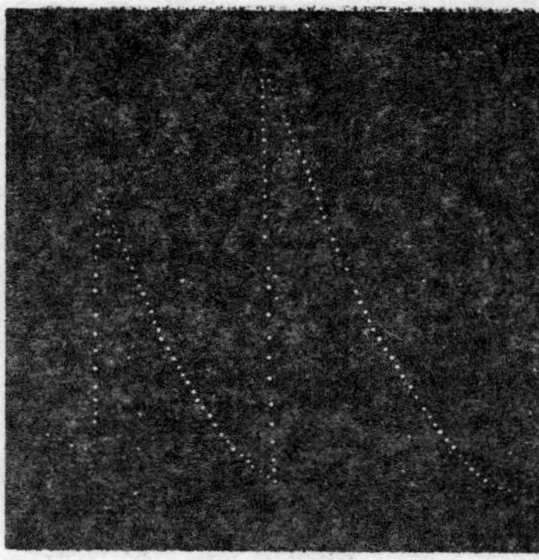

Fig. 127.

Fig. 128.

FIG. 127.—Effect of angle of inclination of light on the tropic curvature of pulvinus. The first response is to light at 45° and the second, to 90°. (*Desmodium gyrans*).

FIG. 128.—Series of tropic curvatures of growing bud of *Crinum* to alternate stimulation by light at 45° and 90°.

effect increases with the directive angle. In the present case the ratio of the two effects is 1·6 : 1, which is not very different from the ratio $\frac{\sin 90°}{\sin 45°} = 1{\cdot}4$.

Tropic response of growing organs: Experiment 132.—Similar experiment was carried out with the flower bud of *Crinum*, held vertical. Light was applied alternately at 45° and 90°, in two successive series. The object of this was to make due allowance of possible variation of excitability of the organ during the course of the experiment. I have explained (p. 147), how the excitability of a tissue in a condition slightly below par, is increased by the action of previous stimulation. Series of responses obtained under alternate stimulations at 45° and 90°

enable us to find out, whether any variation of excitability occurred during the course of the experiment and make allowance for it. The records show that stimulation did enhance the excitability of the organ to a small extent. Thus the first stimulation at 45° induced an amplitude of response of 5 mm. ; the second stimulation at 45° *i. e,* the third response of the series, induced a slightly larger response 7 mm. in amplitude. Similarly the two responses at 90° gave an amplitude of 9 mm. and 11 mm. respectively (Fig. 128). Taking the mean value of each pair, the ratio of tropic effects for 90° and 45 is = 10/6 = 1·7 nearly.

EFFECT OF DURATION OF EXPOSURE.

Experiment 133.—The specimen employed for the experiment was a flower bud of *Crinum* in a slightly sub-tonic condition. Successive responses exhibited on this account, a preliminary negative* before the normal positive curvature. The successive durations of exposure were for 1, 2, and

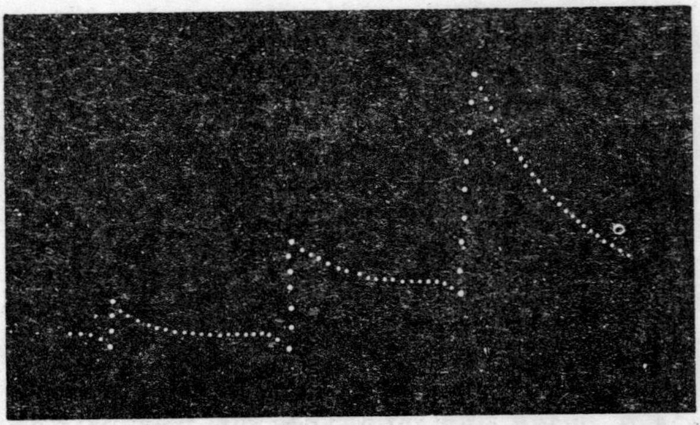

FIG. 129.—Effect of increasing duration of exposure 1 : 2 : 3 minutes, on phototropic curvature. Note preliminary negative response. (*Crinum*).

3 minutes. The amplitudes of responses (Fig. 129) are in the ratio of 1 : 2·5 : 5.

* An explanation of this preliminary effect will be found in the next chapter.

We may now recapitulate the tropic effects of light of increasing intensity, directive angle, and duration of exposure. It has been shown that the tropic effect is enhanced under increasing intensity of light; it is also increased with the angle increasing from grazing to perpendicular incidence. And finally, the tropic effect is enhaneed with the duration of exposure. Taking into consideration the effects of these different factors we arrive at the conclusion that *phototropic effect increases with the quantity of incident light.* It will be shown in the next chapter that strict proportionality of cause and effect holds good in the median range of stimulation, and the slight deviation from this, above and below the median range, is due to the fact that susceptibility for excitation is low at these two regions.

SUMMARY.

Increasing intensity of light induces increasing tropic curvature.

Tropic curvature increases with the directive angle, the effect being approximately proportional to sin θ, where θ is the angle made by the rays with the responding surface.

Tropic curvature also increases with the duration of exposure.

The intensity of induced tropic effect is determined by the quantity of incident light.

XXXII.—THE PHOTOTROPIC CURVE AND ITS CHARACTERISTICS

By

SIR J. C. BOSE.

WHEN a plant organ is subjected to the continued action of unilaterial stimulus of light, it exhibits increasing tropic curvature, which in certain cases reaches a limit; in other instances a reversal takes place, seen in neutralisation, or in the conversion of the positive into negative curvature. I shall in this chapter enter into a detailed study of the phototropic curve, and determine its characteristics.

As the vague terminalogy at present in use has been the source of much confusion, it is necessary here to define clearly the various terms which will be employed in this investigation. It is first of all necessary to distinguish between cause and effect, between external stimulus and the excitation induced by it. As regards stimulus itself I have shown elsewhere* that its effective intensity becomes summated by repetition. This was demonstrated by the two following typical experiments carried out with the pulvinus of *Mimosa*.

(1) The intensity of a single electric shock of intensity of 0·5 unit was found to be ineffective in inducing

* "Irritability of Plants"—p. 54,

excitation; but it became effective on being repeated four times in rapid succession.

(2) The same specimen was next subjected to a feebler stimulus of intensity of 0·1 unit, and it required a repetition of 20 times for the stimulus to become effective.

The total stimulus in the first case was $0{\cdot}5 \times 4 = 2$, and this was found to be the same as $0{\cdot}1 \times 20 = 2$ in the second case. Thus the intensity of stimulus is increased by repetition; in the limiting case where the interval between successive stimulus is zero, the stimulus becomes continuous. Bearing in mind the additive effects of stimulus we see that its effective intensity increases with the *duration* of application. This important conclusion found independent support from the results of Experiment 133 given in the last chapter.

We shall now take up the general question of the characteristics of the phototropic curve, which gives the relation between increasing stimulus and the resulting excitation. As regards stimulus we found that its effectiveness increases with the duration of application. The induced excitation in growing organs may be measured by concomitant retardation of growth caused by stimulus. In the excitation curves which will be presently given, the abscissae represent increasing stimulus and ordinates the resulting excitation. This excitation curve may be obtained by making the plant record on a moving plate its retardation of growth by means of the High Magnification Crescograph. I reproduce below two records of the effects of continuous photic and electric stimulation. The ordinate of the 'excitation curve' (Fig. 130) exhibits increasing incipient contraction (retardation of growth) culminating in an

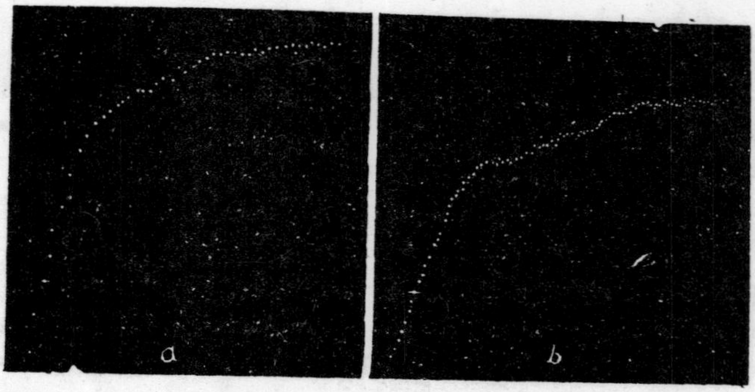

FIG. 130.—Effects of continuous (*a*) electric, and (*b*) photic stimulation on rate of growth. Abscissa represents duration of application of stimulus. Note induced retardation, and arrest of growth.

arrest of growth; the abscissa represents increasing stimulus consequent on increased duration of application. The record shows that the incipient contraction is slight at the first stage; it increases rapidly in the second stage; finally, it declines and reaches a limit. The excitatory reaction is thus not constant throughout the entire curve of excitation, but undergoes very definite and characteristic changes. We shall find similar characteristics in the phototropic curves under unilateral stimulus which will be given presently. The explanation of the similarity is found in the fact that the tropic curvature is also due to incipient contraction or retardation of the rate of growth, which remains confined to the directly stimulated proximal side of the organ.

For facility of explanation of what follows, I shall have to use a new and necessary term, *susceptibility*, to indicate the relation of cause and effect, of stimulus and resulting excitation. *Susceptibility* is thus $= \frac{\text{Excitation}}{\text{Stimulus}}$.

Different organs of plants exhibit unequal susceptibilities; some undergo excitation under feeble stimulus, while others require more intense stimulus to induce excitation. But even in an identical organ the susceptibility undergoes, as we have seen, a characteristic variation, being feeble at the beginning of the excitation curve, considerable in the middle, and becoming feeble once more towards the end of the curve. The most difficult problem that faces us is an explanation of this characteristic difference in different parts of the tropic curve.

GENERAL CONSIDERATIONS.

Before entering into the fuller consideration of the subject, it will be helpful to form some mental picture of the phenomena of excitation, however inadequate it may be. The excitation is admitted to be due to the molecular upset induced by the shock of stimulus*; the increased excitation results from increasing molecular upset brought on by enhanced stimulus. The condition of molecular upset or excitation may be detected from the record of any one of the several concomitant changes, such as the change of form, (contraction or expansion) or change of electric condition (galvanometric negativity or positivity). These means of investigation are not in principle different from those we employ in the detection of molecular distortion in inorganic matter under increasing intensities of an external force.

THE CHARACTERISTIC CURVE.

Thus the molecular upset and rearrangement, in a magnetic substance under increasing magnetising force are

* I shall use the term *stimulus* in preference to *stimulation*, for the latter is often taken in the sense of the resulting excitation.

inferred from the curve obtained by means of appropriate magnetometric or galvanometric methods. I reproduce the characteristic curve of iron (Fig. 131) in which the abscissa represents increasing magnetising force, and the ordinate, the induced magnetisation. This characteristic curve, giving the relation of cause and effect, will be found to be highly suggestive as regards the similar characteristic curve which gives the relation between increasing stimulus and the resulting enhanced tropic effect in vegetable tissues. The parallelism will be found to be very striking.

Inspection of figure 131 shows that, broadly speaking, the curve of magnetisation may be divided into four parts. In the first part, under feeble magnetic force, the slope of the curve is very small; later, in the second part, as the force increases, the curve becomes very steep; in the third part the slope of the curve remains fairly constant; and finally in the fourth part, the curve rounds off and the rate of ascent again becomes very small. The *susceptibility* for induced magnetisation is thus very feeble at the beginning; under increasing force, in the second stage, the susceptibility becomes greatly enhanced; in the third stage, the susceptibility remains approximately constant; and in the fourth stage it becomes diminished. We shall presently find that the susceptibility for excitation also undergoes a similar variation at the four different stages of stimulation.

CHARACTERISTICS OF SIMPLE PHOTOTROPIC CURVE.

I have shown (Fig. 130) the relation between the stimulus and the resulting excitation, the latter being determined from the diminution of the rate of growth. Under unilateral action of light, the excitatory contraction gives rise to tropic curvature. We may thus obtain the characteristic excitation curve, by making the plant organ record its tropic movement under continuous action of light applied on one side of the organ.

Experiment 134.—I give below the characteristic curve of excitation (Fig. 132) of the pulvinus of *Erythrina indica*, traced by the plant itself, and exactly reproduced by photo-mechanical process. A parallel beam of light from a Nernst

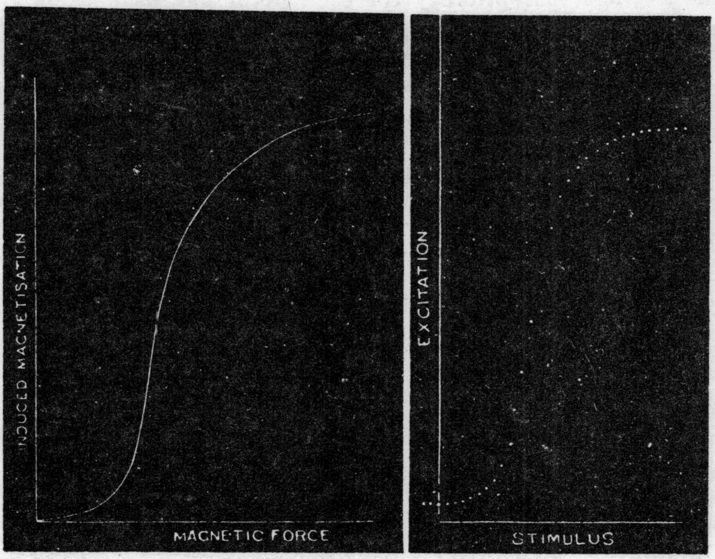

Fig. 131. Fig. 132.

FIG. 131.—Characteristic curve of iron under increasing magnetising force (After Ewing).

FIG. 132.—Simple characteristic curve of phototropic reaction. Excitation increases slowly in the first part and rapidly in the second; it is uniform in the third, and undergoes decline in the fourth part (*Erythrina indica*).

lamp was thrown on the upper leaf of the pulvinus, and the increasing positive curvature was recorded on a smoked glass plate which was moved at an uniform rate. The successive dots are at intervals of 20 seconds; the horizontal distances between successive dots are equal, and represent equal increments of stimulus; the vertical distances between successive dots represent the corresponding increments of excitation. The gradient at any point of the curve—increment of excitation divided by increment of stimulus—gives the susceptibility for excitation at that point. The following table will show how the susceptibility for excitation undergoes variation through the entire range of stimulus. The average susceptibility for each point has been calculated from the data furnished by the curve.

TABLE XXX.—SHOWING THE VARIATION OF SUSCEPTIBILITY FOR EXCITATION AT DIFFERENT POINTS OF THE TROPIC CURVE.

Successive points in the curve.	Susceptibility for excitation.	Successive points in the curve.	Susceptibility for excitation.
1	0	14	6·6
2	0·187	15	4·4
3	0·44	16	2·5
4	0·625	17	1·87
5	0·875	18	1·5
6	1·25	19	1·12
7	1·87	20	0·937
8	3·12	21	0·75
9	5·0	22	0·562
10	6·25	23	0·375
11	8·75	24	0·25
12	8·87	25	0·187
13	8·12	26	0·062

The induced excitation is seen to be increased very gradually from the zero point of susceptibility, known as the latent period at which no excitation takes place. In the second part of the excitation curve, the rate of increase is very rapid; the maximum rate is nearly reached at point 11 of the curve and remains fairly constant for a time. This is the median range where equal increment of stimulus induces equal increment of excitation. The susceptibility for excitation then falls rapidly, and increase of stimulus induces no further increase of tropic curvature.

The maximum tropic curvature was, in the present case, reached in the course of nine minutes. The attainment of this maximum depends on the excitability of the tissue, and the intensity of incident stimulus. The characteristics that have been described are not confined to the phototropic curve but exhibited by tropic curves in general. Similar characteristics have been found in the curve for electric stimulus (Fig. 130a), and will also be met with in the curve for geotropic stimulus (Fig. 161).

I may here refer incidentally to the three types of responses exhibited by an organ to successive stimuli of uniform intensity; these appear to correspond to the three different regions of tropic curve; in the first stage, the plant exhibits a tendency to exhibit a 'staircase' increase of response; in the intermediate stage, the response is uniform; and in the last stage, the responses show a fatigue' decline.

For purpose of simplicity, I first selected the simple type of phototropic curve, where the specimen employed was in a favourable tonic condition, and the stimulus was, from the beginning, above the minimal. Transverse conduction, which induces neutralisation or reversal into negative, was moreover absent in the specimen. I shall now take up the more complex cases: (1) where the condition of the specimen is slightly sub-tonic, (2) where the stimulus is gradually increased from the *sub-minimal*, and (3) where the specimen possesses the power of transverse conduction.

EFFECT OF SUB-MINIMAL STIMULUS.

It is unfortunate that the terms in general use for description of effective stimulus should be so very indefinite. A stimulus which is just sufficient to evoke excitatory *contraction* is termed the minimal, stimulus below the threshold being tacitly regarded as ineffective. The employment of sensitive recorders has, however, enabled me to discover the important fact that stimulus below the minimal, though ineffective in inducing excitatory *contraction*, is not below the threshold of perception. The plant not merely perceives such stimulus, but also responds to it in a definite way, by *expansion* instead of *contraction*. I shall designate the stimulus below the minimal, as the *sub-minimal*. There is a critical point, which demarcates the sub-minimal stimulus with its expansive reaction from the minimal with its responsive contraction.

The *critical* stimulus varies in different species of plants. Thus the same intensity of light which induces a retardation of growth in one species of plants will enhance the rate of growth in another. Again, the critical point will vary with the *tonic* level of the same organ; in an optimum condition of the tissue, a relatively feeble stimulus will be sufficient to evoke excitatory contraction; the critical point is therefore low for tissues in tonic condition which may be described as *above par*. In a sub-tonic condition, on the other hand, strong and long continued stimulation will be necessary to induce the excitatory reaction. The critical point is therefore high, for tissues in a condition *below par*. Stimulus below the critical point will here induce the opposite physiological reaction, *i.e.*, expansion. The physico-chemical reactions underlying these opposite physiological responses have, for convenience, been distinguished as the "A" and "D" change (pp. 143, 223). The assimilatory 'building up', A change, is associated with an increase of potential energy of the system; the dissimilatory 'break down', D change, on the other hand, is attended by a run-down of energy.

Stimulus was shown (p. 225) to give rise to *both* these reactions, though the A effect is, generally speaking, masked by the predominant D effect. The "A" change is quicker in initiation, while the "D" effect developes later; again the "A" effect under moderate stimulation may persist longer. Thus owing to the difference in their time-relations the A effect is capable of being unmasked at the onset of stimulus or on its sudden cessation. For the detection of the relatively feeble expansive A effect, a special recorder is required which combines lightness with high power of magnification. The earlier expansive reaction and acceleration of rate of growth, followed by normal retardation, are often found in the response of growing organs. The corresponding effect of unilateral stimulation, even when direct, is a transient expansion at the proximal side, inducing a convexity of that side and movement away from stimulus (negative curvature); this is followed by contraction and concavity with normal positive curvature. The interval between the A and D effects is increased with increasing sub-tonicity of the specimen. But it nearly vanishes when the excitability of the specimen is high, and the two opposite reactions succeed each other too quickly for the preliminary A reaction to become evident. It is probable that in such a case the conflict between the two opposite reactions prolongs the latent period. But in other instances a preliminary expansive response is found to herald the more pronounced contractile response. Example of this is seen in figure 129 given in page 344.

The A effect was detected in the records referred to above by its earlier appearance. Its longer persistence, after moderate stimulation, is also to be found on the cessation of moderate stimulation. This was seen in the *acceleration* of growth which was the after-effect of stimulation (Figs. 104, 115). The presence of two conflicting physiological reactions is also made evident on sudden cessation of long continued stimulation. This particular phenomenon of "overshooting" will be more fully dealt with in a subsequent chapter.

Owing to the difference in the time relations of the two opposing activities, A and D, a phase difference often arises in their respective maxima. It is probably on this account that rhythmic tissues originally at standstill, exhibit under continued stimulation a periodic up and down-movement, which persists even on the cessation of the stimulus. The persistence of after-oscillation depends, moreover, on the intensity and duration of previous stimulation.*

The facts given above cannot be explained by the prevalent theory that stimulus acts merely as a releasing agent, to set free energy which had been previously stored up by the organism, like the pull of a trigger causing explosion of a charged cartridge. It is true that in a highly excitable tissue, the external work performed and the run down of energy are disproportionately greater than the energy of stimulus that induces it. But in a sub-tonic tissue, stimulus induces an effect which is precisely the opposite ; instead of a depletion, there is an enhancement of potential energy of the system. Thus the responding leaf instead of undergoing a fall becomes erected ; growing organs similarly exhibit a 'building up' and an acceleration of rate of growth, in contrast with the usual 'break down' and depression of the rate. It is obvious that these new facts relating to the action of stimulus necessitate a theory more comprehensive and satisfactory than the one which has been in vogue.

THE COMPLETE PHOTOTROPIC CURVE.

I have explained the characteristics of the simple phototropic curve in which the tropic curvature, on account of the favourable tonic condition and strong intensity of

* "Plant Response"—p. 293, etc.

incident light, was positive from the beginning, and in which the curvature reached a maximum beyond which there was no subsequent reversal. If the intensity of the stimulus be feeble or moderate, the quantity of light incident on the responding organ at the beginning may fall below the critical value, and thus act as a sub-minimal stimulus. This induces as we have seen (p. 344) a negative tropic curvature; continued action of stimulus, however, converts the preliminary negative into the usual positive. The preliminary negative curvature may be detected by the use of a moderately sensitive recorder with a magnification of about 30 times. It is comparatively easy to obtain the preliminary negative response in specimens which are in a slightly sub-tonic condition.

Semi-conducting tissues exhibit under continued stimulation, a neutralisation and reversal into negative (p. 331). Since this reversal into negative usually takes place under prolonged exposure to exceedingly strong light, it is difficult to obtain in a single curve all the different phases of transformation. I have, however, been fortunate in obtaining a complete phototropic curve which exhibits in a single specimen all the characteristic changes from a preliminary negative to positive and subsequent reversal to negative. I shall describe two such typical curves obtained with the terminal leaflet of *Desmodium gyrans* and the growing seedling of *Zea Mays*.

Complete phototropic curve of a pulvinated organ: Experiment 135.—A continuous record was taken of the action of light of a 50 c. p. incandescent lamp, applied on the upper half of the pulvinus of the terminal leaflet of *Desmodium gyrans*. This gave rise: (1) to a negative curvature (due to sub-minimal stimulus) which lasted for 3 minutes. The curve then proceeded upwards, at first slowly, then rapidly; it then rounded off, and reached a maximum positive value in the course of 18 minutes; after this the curve underwent a reversal, and complete neutralisation occurred after a further period of 24 minutes (Fig. 133). Beyond this the induced curvature is negative.

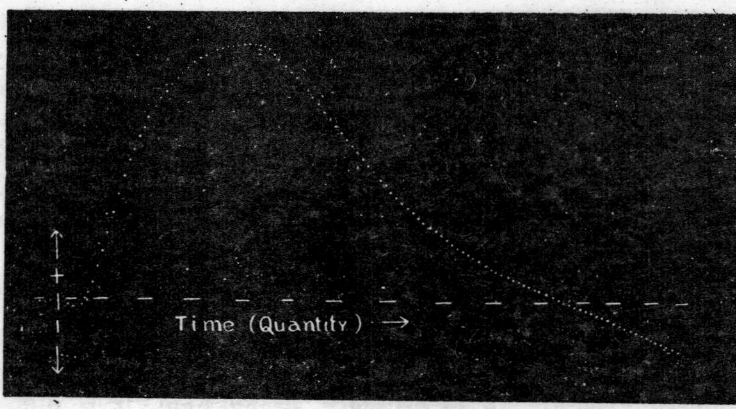

FIG. 133.—Complete phototropic curve given by pulvinated Eq. organ. Positive curvature above, and negative curvature below the horizontal zero line. Preliminary negative phase of response due to sub-minimal stimulus. The positive increases, attains a maximum, and undergoes a reversal. Successive dots at intervals of 30 seconds. Abscissa represents duration of exposure and quantity of incident light. (Terminal leaflet of *Desmodium gyrans.*)

Complete phototropic curve of growing organs: Experiment 136.—I obtained very similar effects by subjecting the seedling of *Zea Mays* to unilateral light from an arc lamp for two hours. The characteristic of this curve is similar to that given by the terminal leaflet of *Desmodium gyrans.* At the first stage, the sub-minimal stimulation is seen to induce a negative curvature, transformed into positive after an interval of 10 minutes. The maximum positive curvature is reached after 50 minutes, and neutralisation completed in a further period of 43 minutes

(Fig. 134). After this the response became transformed into negative.

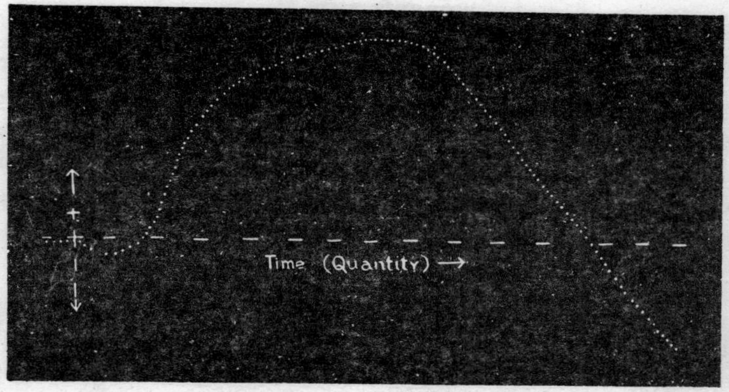

FIG. 134.—Complete phototropic curve of a growing organ (*Zea Mays*).

In a complete phototropic curve we may thus distinguish 4 distinct stages :—

(1) The stage of sub-minimal stimulation.
(2) The stage of increasing positive curvature culminating in a maximum.
(3) The stage of neutralisation.
(4) The stage of complete reversal into negative.

The curve thus crosses the zero line of the abscissa twice ; the first crossing takes places *upwards* at the critical point of stimulation which demarcates the sub-minimal from the minimal. The second crossing downwards occurs beyond the point of complete neutralisation.

In a tissue in which transverse conductivity is absent, and the stimulus applied from the beginning is above the minimal, the simple tropic curve is confined to the second stage (see Fig. 132).

WEBER'S LAW.

If we neglect the preliminary negative portion under sub-minimal stimulus, the curve of excitation under increasing photic stimulation obeys what is known as Weber's law. This is equally true of modes of stimulation other than that of light as is seen in figure 130 of the contractile effect of continued electric stimulus on growth; the excitatory effect is also seen to reach a limit.

Weber's law is applicable for a limited range of stimulation. For the quantitative relation fails in the region of sub-minimal stimulus, where the physiological reaction is *qualitatively* different, namely expansion instead of contraction. This holds good even in the case of animal tissues, for here also my recent experiments show that two opposite reactions—expansion and contraction—take place under stimulus, and that a very feeble stimulus tends to induce expansion instead of contraction. The responsive reaction of a kitten under gentle caressing strokes must be *qualitatively* different from that of a blow. The psychological effects under the two treatments evidently differ qualitatively rather than quantitatively.*

SUMMARY.

The excitation curve exhibits a slow ascent in the first part; in the second part the gradient is steep, indicating rapid rise in excitation; in the third part it is

* "It has been argued by James that the feeling does not cause, but is caused by the bodily expression . . . . Münsterberg concludes that the feeling of agreeableness is the mental accompaniment and outcome of reflexly produced movements of extension, and disagreeableness of the movement of flexion." Schäfer—Text Book of Physiology, Vol. II, p. 975 (1900).

uniform ; and in the last part the curve rounds off and the rate of ascent becomes very small.

The susceptibility for excitation is feeble at the beginning ; it increases very rapidly with increasing stimulus ; finally it undergoes a fall, increase of stimulus inducing no further enhancement of excitation.

In a complete phototropic curve the first part is negative ; this is due to the physiological expansion induced by sub-minimal stimulus. The curve then crosses the abscissa upwards, and the positive curvature reaches a maximum. This is followed by neutralisation and reversal into negative ; the curve crosses the zero line and proceeds in the negative direction.

Weber's law is not applicable for the entire range of stimulation. The quantitative relation fails in the region of sub-minimal stimulus, where the physiological reaction is *qualitatively* different.

XXXIII.—THE TRANSMITTED EFFECT OF PHOTIC STIMULATION

By

SIR J. C. BOSE,

Assisted by

JYOTIPRAKASH SIRCAR, M.B.

PLANT organs exhibit, as we have already seen, a heliotropic curvature under direct stimulation. Still more interesting is the transmitted effect of light giving rise to a curvature. Thus if the tip of the seedling of wheat be exposed to light, the excitation is transmitted lower down into the region which acts as the responding organ. Growth is very active in this particular zone, and the change of growth, induced by the transmitted effect of stimulus, brings about a curvature by which the tip of the seedling bends towards light. The seedling thus appears to be differentiated into three physiological zones subserving three different functions. The tip is the perceptive zone, the intervening distance between the tip and the growing region is the zone of conduction, and the growing region is the responsive zone. These differentiations are shown in a striking manner by certain Paniceae, *Setaria* for example. In this seedling the tapering sheathing leaf or cotyledon is about 5 mm. in length, and it is the upper part of the cotyledon that is most sensitive to light. Below the sheathing leaf is a narrow length which will be distinguished as the hypocotyl, and where growth is very active. The apex of the leaf perceives the stimulus, and the effect is transmitted to the hypocotyl, which responds by becoming curved so that the seedling bends towards light.

It is necessary here to make special reference to the confusion that arises from want of precision in the use of the term stimulus, used indifferently to denote both the cause and the resulting effect. An external agent, say light, causes certain excitatory change in the tissue, and we refer to the agent which induces it, as the *stimulus.* Thus in the instance cited above, light is the *stimulus*, and it is the stimulus-effect that is transmitted to a distance. But in physiological literature no distinction is made between the stimulus and its effect, hence arises frequent use of the phrase 'transmission of stimulus'. It is obvious that it is not light but its effect that is transmitted.

Such want of precision in the use of the term stimulus would not have seriously affected the truth about the description of facts, had the transmitted effect been only of one kind. In a nerve-and-muscle preparation, the velocity of transmission of excitation is so great, that it completely masks the positive impulse (assuming the existence of such an impulse). The effect of indirect stimulation is, therefore, the same as that of direct stimulation. Any indefiniteness in the use of the term stimulus for its transmitted effect does not, in animal physiology, seriously militate against the observed facts. But lack of precision in the employment of the term in plant physiology leads to hopeless confusion. For owing to the semi-conducting nature of vegetable tissue, the transmitted effect is not of a definite sign, but may be positive or negative; in the first case, the response is by expansion, in the latter, by contraction. Thus the transmitted effect will be very different in the two cases, according as the intervening tissue is a good or a bad conductor. These facts accentuate the urgent necessity of revision of our existing terminology.

I have shown that the effects of other forms of stimuli are also transmitted from the perceptive to the responding region along the intervening path of conduction. Thus the petiole of *Mimosa* perceive any form of stimulus applied to it, and the induced excitation is conducted to the distant pulvinus to evoke the familiar responsive fall of the leaf. The pulvinus, moreover, perceives and responds to direct stimulation. In a nerve-and-muscle preparation the responding muscle is alike perceptive and responsive.

But in *Setaria* we meet with certain characteristics of reaction which are quite inexplicable. Thus if

"the seedling be illuminated on one side, a sharp heliotropic curving takes place at the apex of hypocotyl. The curvature makes itself apparent only if the cotyledon be illuminated from one side whether the hypocotyl be exposed to light or not. If the cotyledon be shaded and the light be permitted to fall on one side of the hypocotyl, no heliotropic curving takes place. Hence we may conclude that it is only the cotyledon that is sensitive to the light stimulus, and it is only the hypocotyl which can carry out the movement. The excitation which the light effects in the cotyledon must be transmitted to the hypocotyl and curvature takes place only from such a transmitted excitation. We have thus in this case a definite organ for the perception of the stimulus of light, viz., the cotyledon, and as Rothert has shown, it is more specially the apex of that organ that is the sensitive part: on the other hand, the motile organ, the hypocotyl, is some distance away from the sensitive organ, and in it the power of perception is entirely absent. From the behaviour of these organs we may draw the further conclusion that perception and heliotropic excitation are two distinct phenomena, which depend on different properties of the protoplasm and which are independent of each other . . . We may, therefore, conclude from this experiment that these two types of excitation are fundamentally distinct processes, for it is only after indirect or transmitted and not after direct excitation that a reaction occurs in the case of the seedlings of the Paniceae".*

The noteworthy deductions on the above facts are :—

(1) That the motile organ in *Setaria* is totally devoid of perception, since direct action of light induces no effect.

(2) That perception and heliotropic excitation are two distinct phenomena, which depend on different properties of the protoplasm, and which are independent of each other.

Though the conclusions thus arrived at appear to follow from the facts that have been observed, yet it is difficult to accept the inference, that a responding organ should be totally devoid of the power of perception, and that excitation and perception are to be regarded as dependent on different properties of protoplasm. It therefore appeared necessary to re-investigate the subject of the perceptive power of the cotyledon, and the responding characteristics of the hypocotyl.

The criterion employed for test of perception is the movement induced in response to stimulus. The responsive *mechanical* movement is rendered possible only by the contractility of the organ, and mechanical and anatomical facilities offered by it for unhampered movement. The

* Jost —*Ibid* —p. 468.

petiole of *Mimosa* when locally stimulated does not itself exhibit any movement. The fortunate circumstance of the presence of a motile pulvinus in the neighbourhood enables us to recognise the perceptive power of the petiole, since it transmits an impulse which causes the fall of the leaf. There is no motile pulvinus in ordinary leaves, and stimulation of the petiole gives rise to no direct or transmitted motile reaction; from this we are apt to draw the inference that the petiole of ordinary leaves are devoid of perception. This conclusion is, however, erroneous, since under stimulus the petiole exhibits the electric response characteristic of excitation. Moreover my electric investigations have shown that every living tissue not only perceives but also responds to stimulation.* Hence considerable doubt may be entertained as regards the supposed absence of perception in the hypocotyl of *Setaria*.

I shall in the present paper describe my investigations on the mechanical response of *Setaria* under direct and indirect stimulation which will be given in the following order:—

(1) The response to unilateral stimulation of the tip of the seedling.

(2) The response of growing hypocotyl to direct stimulation.

(3) Summated effects of direct and indirect stimulation.

EXPERIMENTAL ARRANGEMENTS.

The Recorder.—The pull exerted by the tropic curvature of the seedling is very feeble; it was therefore necessary to construct a very light and nearly balanced

* "Response in the Living and Non-Living"—p. 17.

recording lever. A long glass fibre is supported by lateral pivots on jewel bearings. The seedling is attached to the short arm of the lever by means of a cocoon thread. The recording plate oscillates to and fro once in a minute; the successive dots give therefore the time relations of the responsive movement. The positive curvature towards light is recorded as an up-curve, the negative curvature being represented by a down-curve.

Arrangement for local stimulation by light.—The device of placing tin foil caps on the tip employed by some observers labours under the disadvantage, that it causes mechanical irritation of the sensitive tip. The appliance seen in figure 135 is free from this objection and offers

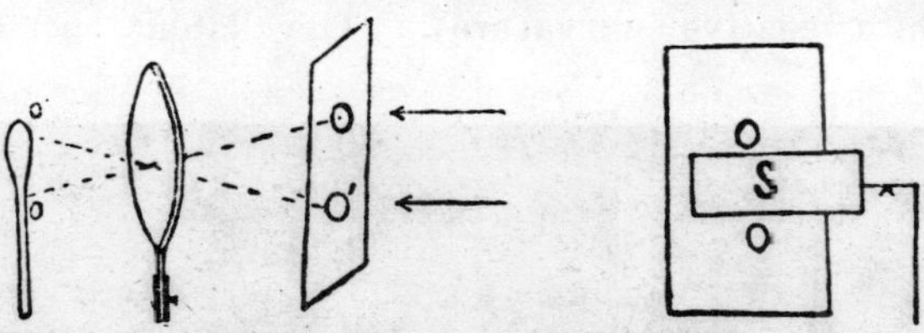

FIG. 135.—Arrangement for local application of light to the tip and the growing region. O, O[1], apertures on a metallic screen. Light is focussed by a lens on the tip, and on the growing region at o, o.[1] Figure to the right shows front view of the shutter resting on a pivot and worked by string, T.

many advantages. A metallic screen has two holes O and O[1]; these apertures are illuminated by a parallel beam of light from an arc lamp. A lens focusses the light from O, on the hypocotyl, and that from O[1], on the tip of the cotyledon. A rectangular pivoted shutter S, lies between the apertures O and O[1]. In the intermediate position of the shutter, light acts on both the tip and the growing region. The shutter is tilted up by a pull on the thread T, thus cutting off light from the growing region; release of the thread cuts off light from the tip. Thus by proper manipulation of the shutter, the tip or the growing hypocotyl, or both of them, may be subjected to the stimulus of light. The experiment was carried out in a dark room, special precaution being taken that light was screened off from the plant except at points of localised stimulation.

EFFECT OF LIGHT AT THE TIP OF THE ORGAN.

Experiment 137.—If the tip of the seedling of *Setaria* be illuminated on one side, it is found that a *positive* curvature (*i.e.*, towards light) is induced in the course of an hour or more. But in obtaining record of the seedling by unilateral stimulation of the tip, I found that the immediate response was not towards, but away from light (negative curvature). The latent period was

FIG. 136.—Response of seedling of *Setaria* to unilateral stimulation of the tip applied at dotted arrow.

Note preliminary negative curvature reversed later into positive.

about 30 seconds and the negative movement continued to increase for 25 minutes (Fig. 136). This result, hitherto unsuspected, is not so anomalous as would appear at first sight. Indirect stimulus, unilaterally applied, has been shown to give rise to two impulses: a quicker positive and a slower excitatory negative. The former induces a convexity on the same side, and a movement away from stimulus (negative curvature); the excitatory negative, on the other hand, is conducted slowly and induces contraction and concavity, and a movement towards the stimulus (positive curvature). In semi-conducting or non-conducting tissues, the excitatory negative is weakened to extinction during transit, and the positive reaction with negative curvature persists as the initial and final effect.

But in *Setaria* the excitatory negative impulse is transmitted along the parenchyma which is moderately conducting; the speed of transmission of heliotropic excitation is, according to Pfeffer, one or two mm. in five minutes or about 0·4 mm. per minute. Thus under the continued action of light, the excitatory impulse will reach the growing region, and by its predominant reaction neutralise and reverse the previous negative curvature.

Inspection of figure 136 shows that this is what actually took place; the intervening distance between the tip of the cotyledon and the growing region in hypocotyl was about 20 mm., and the beginning of reversal from negative to positive curvature occurred 29 minutes after application of light. The velocity of transmission of excitatory impulse under strong light is thus 0·7 mm. per minute. The positive curvature continued to increase for a very long time and became comparatively large. This is for two reasons: (1) because the sensibility of the tip of the cotyledon is very great, and (2) because the positive curvature induced by longitudinally transmitted excitation is not neutralised by transverse conduction (see below).

RESPONSE TO UNILATERAL STIMULUS IN THE GROWING REGION.

Experiment 138.—The growing region of the hypocotyl of *Setaria* is supposed to be totally devoid of the power of perception. In order to subject the question to experimental test, I applied unilateral light on the growing region of the same specimen, after it had recovered from the effect of previous stimulation. The response now obtained was vigorous and was *ab-initio* positive. Direct stimulus has thus induced the normal effect of contraction and concavity of the excited side. The belief that the hypocotyl of *Setaria* is incapable of perceiving stimulus is thus without any foundation. The further experiment

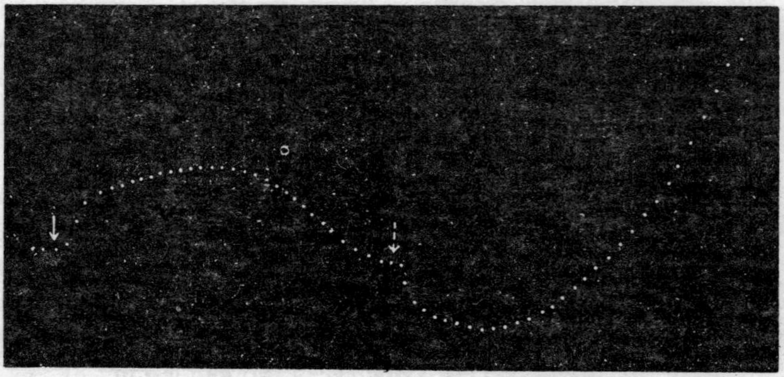

FIG. 137.—Effect of application of light to the growing hypocotyl at arrow induced positive phototropic curvature followed by neutralisation. Application of indirect stimulus at dotted arrow on the tip gave rise at first to negative, subsequently to positive curvature. (Seedling of *Setaria*).

which I shall presently describe will, however, offer an explanation of the prevailing error. On continuing the action of unilateral light, the positive curvature after attaining a maximum in the course of 15 minutes, underwent a diminution and final neutralisation (Fig. 137). On account of this neutralisation the seedling became erect after an exposure of 30 minutes; in contrast with this is the increasing positive curvature under unilateral illumination of the tip (Fig. 136) which continues for several hours. The explanation of this neutralisation under direct stimulation of the growing region is found in the fact that transverse conduction of excitation induces contraction at the distal side of the organ and thus nullifies the positive curvature. The seeming absence of tropic effect under direct stimulation is thus not due to want of perception, but to balanced antagonistic reactions on opposite sides of the organ.

EFFECT OF SIMULTANEOUS STIMULATION OF THE TIP AND THE HYPOCOTYL.

Though stimulation of the hypocotyl results in neutralisation, yet the illumination of one side of the organ including the tip and hypocotyl is found to give rise to positive curvature. This will be understood from the following experiment.

After the neutralisation in the last experiment light was also applied to the tip from the right side at the dotted arrow (Fig. 137). The record shows that this gave rise at first to a negative curvature (away from light); under the continued action of light, however, the negative was subsequently reversed to a positive curvature, towards light. Inspection of the curve shows another interesting fact. The positive curvature induced by direct stimulation is very much less than that brought out by indirect stimulation. This is due to two reasons: (1) the sensitiveness of the tip of the organ is, as is well known, greater than that of the hypocotyl, (2) the positive curvature under direct stimulation cannot proceed very far, since it is neutralised by transverse conduction of excitation.

It will be seen from the above that the illumination of the tip practically inhibits the neutralisation and thus restores the normal positive curvature. The question now arises as to how this particular inhibition is brought about.

ALGEBRAICAL SUMMATION OF THE EFFECTS OF DIRECT AND INDIRECT STIMULATIONS.

An instance of inhibition, though of a different kind, was noticed in the response of the tendril of *Passiflora* (p. 296); the under side of the organ is highly sensitive, while the upper side is almost insensitive. Stimulation of the under side of the tendril induces a marked curvature, but simultaneous stimulation of the diametrically opposite side inhibits the response. This neutralisation could not be due to the antagonistic contraction of the upper side since the irritability of that side is very slight. I have shown that the inhibition results from the two antagonistic reactions, contraction at the proximal side due to direct stimulation and expansion caused by the positive impulse from the indirectly stimulated distal side.

We have in the above an algebraical summation of the effects of direct and indirect stimulations. The longitudinally transmitted effect of indirect stimulus in *Setaria* may, likewise, be summated with the effect of direct stimulus. The phenomenon of algebraical summation is demonstrated in a very striking and convincing manner in the following experiment, which I have been successful in devising.

Experiment 139.—I have explained, (Expt. 126) that unilateral application of stimulus of light on the upper half of the responding pulvinus of *Mimosa* induces an up or positive curvature, followed by a neutralisation and even a reversal into negative, the last two effects being brought about by transverse conduction of excitation to the distal side. When the incident light is of moderate intensity, the transmitted excitation only suffices to induce neutralisation

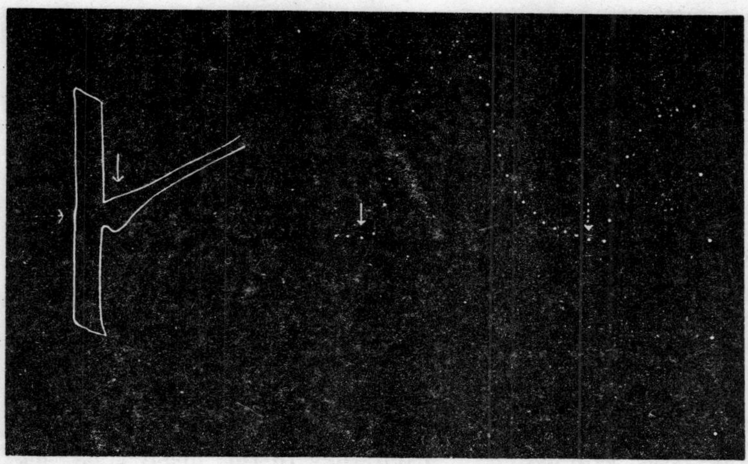

FIG. 138.—(*a*) Diagrammatic representation of direct application of light (↓) on the pulvinus and the indirect application on the stem (—→) (*b*) Record of effect of direct stimulus, positive curvature followed by neutralisation. Superposition of the positive reaction of indirect stimulus induces erectile up-response followed by down movement due to transmitted excitatory impulse (*Mimosa*).

without further reversal into negative; while in this state of balanced neutralisation let us apply indirect stimulus by throwing light on the stem at a point directly opposite to the leaf (Fig. 138).

Two different impulses are thus initiated from the effect of indirect stimulus. In the present case the positive reached the responding pulvinus after 30 seconds and induced an erectile movement of the leaf; the excitatory negative impulse reached the organ 4 minutes later and caused a rapid fall of the leaf. The record (Fig. 138) shows further that the previous action of direct stimulus which brought about neutralisation, does not interfere with the effects of indirect stimulus. The individual effects of direct and indirect stimulus are practically independent of each other; hence their joint effects exhibit algebraical summation.

We are now in a position to have a complete understanding of the characteristic response of Paniceae to transmitted phototropic excitation.

(1) Local stimulation of the tip gives rise to two impulses, positive and negative. The former induces a transient negative movement (away from light); the latter causes a permanent and increasing positive curvature towards light.

(2) Local stimulation of the growing hypocotyl gives rise to positive curvature, subsequently neutralised by the transverse conduction of excitation to the distal side. The absence of tropic effect in the growing region is thus due not to lack of power of perception, but to balanced antagonistic reactions of two opposite sides of the organ.

(3) The effects of direct and indirect stimulations are independent of each other; hence, on simultaneous stimulations of the tip and the growing hypocotyl, the effects of indirect stimulus are algebraically summated with the effect of direct stimulus (neutralisation). The indirect stimulation of the tip on the right side gives rise to two impulses, of which the expansive positive reached the right side of the responding region earlier, inducing

convexity and movement away from stimulus (negative curvature). This is diagrammatically shown in Fig. 139

Had the intervening tissue been non-conducting, the slow excitatory negative impulse would have failed to reach the responding region, and the negative curvature induced by the positive impulse would prove to be the initial as well as the final effect. In the case of *Setaria*, however, the excitatory impulse reaches the right side of the organ after the positive impulse ; the final effect is therefore an induced concavity and positive curvature (movement towards stimulus).

FIG. 139.—Diagrammatic representation of the effects of direct and indirect stimulus on the response of *Setaria*. Direct stimulation, represented by thick arrow gives rise to antagonistic concavities of opposite sides of responding hypocotyl, resulting in neutralisation.

Indirect stimulus represented by dotted arrow gives rise to two impulses, the quick positive impulse represented by a circle, and the slower negative impulse represented by crescent (concave).

The results given above enable us to draw the following generalisations :—

1. In an organ, the tip of which is highly excitable, the balanced state of neutralisation, induced by direct stimulation of the responding region, is upset in two different ways by two impulses generated in consequence of indirect stimulation at the tip. Hence arises two types of resultant response :—

Type A.—If the intervening tissue be semi-conducting, the positive impulse alone will reach the growing region and induce convexity of the same side of the organ giving rise to a *negative* curvature.

Type B.—If the intervening tissue be conducting the transmission of the excitatory impulse will finally give rise to a *positive* curvature.

Type B is exemplified by the seedling of *Setaria* where the transmission of excitatory impulse from the tip upsets the neutral balance and induces the final positive curvature.

Example of type A is found in the negative phototropism of the root of *Sinapis.*

Negative phototropism of root of Sinapis: *Experiment 140.*—For investigation of the negative phototropism of the root of *Sinapis nigra* I took record of its movement under unilateral action of light by means of a Recording Microscope, devised for the purpose.* When the root-tip alone was stimulated by unilateral light, the root moved away from the source of light. This was due to the longitudinal transmission of positive impulse to the growing region at some distance from the tip. The intervening distance between the tip and the growing region is practically non-conducting, hence the excitatory impulse could not be conducted from the tip. After a period of rest in darkness, I next took record of its movement under direct unilateral illumination of the growing region; the result was at first a positive movement; but this, on account of transverse conduction of excitation under continued stimulation, underwent a neutralisation and slight reversal. In taking a third record, in which both the tip and growing region were simultaneously subjected to unilateral stimulation of light, I found that a resultant responsive movement was induced which was away from light.

Thus in the root of *Sinapis*, the expansive effect of indirect stimulation of the tip is superposed on that of direct stimulation of the growing region (neutral or slightly

* "Plant Response"—p. 601.

negative). The final result is thus a movement away from light or a *negative* phototropic curvature.

SUMMARY.

The effect induced by stimulus of light is transmitted to a distance, in a manner precisely the same as in other modes of stimulation.

In the Paniceae, the local unilateral stimulation of the tip of the cotyledon induces positive curvature in the growing hypocotyl, at some distance from the tip. This is due to transmitted excitatory effect of indirect stimulation; the earlier positive impulse induces a preliminary negative curvature, which is reversed later by the excitatory negative impulse into positive curvature.

Contrary to generally accepted view the hypocotyl not only perceives but responds to light. The positive curvature induced by direct stimulation is, however, neutralised by transverse conduction of excitation.

The effects of direct and indirect stimulus are independent of each other; the final effect is determined by their algebraical summation.

XXXIV.—ON PHOTONASTIC CURVATURES

By

SIR J. C. BOSE,

Assisted by

GURUPRASANNA DAS.

PHOTOTROPIC response, positive or negative, is determined by the directive action of light. But photonastic reaction is supposed to belong to a different class of phenomenon, where the movement is independent of the directive action of light. I shall, however, be able to establish a continuity between the tropic response of a radial and the nastic movement of a dorsiventral organ. The intermediate link is supplied by organs originally radial, but subsequently rendered anisotropic by the unilateral action of stimulus of the environment. In a dorsiventral organ, owing to anatomico-physiological differentiation, the responsive movement is constrained to take place in a direction perpendicular to the plane of separation of the two unequally excitable halves of the organ. Even in such a case, it will be shown, that light does exert a directive action; the direction of movement will further be shown to be distorted by the lateral action of light.

PHOTOTROPIC RESPONSE OF ANISOTROPIC ORGANS.

The different sides of a radial organ, such as the young stem of *Mimosa*, are equally excitable. The response to unilateral light of moderate intensity is therefore positive; owing to equal excitabilities of the two sides the response of the opposite sides are alike. Diffuse stimulation therefore induces no resultant curvature. If, however, the plant is allowed to form a creeping habit, the excitabilities of the dorsal and ventral sides will no longer remain the same. Thus in the creeping stem of *Mimosa* the lower or the shaded side is, generally speaking, found to be the more excitable. In fact such anisotropic stem of *Mimosa* acts somewhat like the pulvinus of the same plant. Diffuse stimulation induces, in both, a concavity of the more excitable lower half with the down movement of the leaf or the stem.

Experiment 141.—I took four creeping stems of *Mimosa* in vigorous condition and tied them in such a manner that their free ends should be vertical. The shaded sides of the four specimens were so turned that each faced a different point of the compass—east, west, north and south. Subjected thus to diffuse stimulus of light from the sky, they all executed curvatures. The specimen whose under side faced the east, became bent towards the east; the same happened to those which faced north, south, and west, that is to say they curved towards the north, south, and west respectively (Fig. 140). The fundamental action by which all these were determined was the induced concavity of the under or normally shaded side, which was the more excitable

I obtained similar results with various other creeping stems.

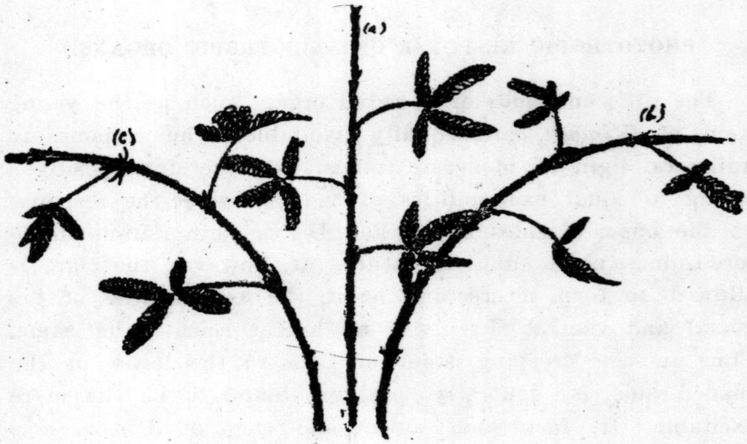

FIG. 140.—Photonastic curvature of creeping stem of *Mimosa pudica:* in the central figure the stem is seen to be vertical: action of diffuse light induced appropriate curvatures by greater contraction and concavity of the more excitable lower or shaded side, as seen in figures to the right (*b*) and left (*c*).

It has been shown that under prolonged unilateral stimulation, excitation becomes internally diffused; this gives rise to an effect similar to that of external diffuse stimulus. Under strong light the shaded side becomes concave, and thus press against the ground or the support; this will be the characteristic response of creeping stems in which the shaded side is the more excitable. The facts given above will probably explain the response of midribs of leaves, of the creeping stem of *Lysimachia*, all of which, in response to the action of strong light acting from above, exhibit concavity of the shaded and more excitable side.

PARA-HELIOTROPISM.

Under strong sunlight, the leaflets of various plants move sometimes upwards, at other times downwards, so as to place the blades of leaflets parallel to incident light. This 'midday sleep' has been termed *para-heliotropism* by Darwin.. It has been thought that para-heliotropic action has nothing to do with the directive action of light, since many leaflets either fold upwards or downwards, irrespective of the direction of incident light. I shall for convenience distinguish the leaflets which fold upwards under light as *positively* para-heliotropic, and those which fold downwards as *negatively* para-heliotropic. This is merely for convenience of description. There is no specific irritabality which distinguishes one from the other.

POSITIVE PARA-HELIOTROPISM.

Para-heliotropic response of Erythrina indica *and of* Clitoria ternatea : *Experiment 142.*—For the purpose of simplicity I have described the type of movement of these leaflets as upwards; but the actual direction in which the leaflets point their apices is towards the sun. Both the plants mentioned here are so remarkably sensitive that the leaflets follow the course of the sun, in such a way that the axis of the cup, formed by the folding leaflets at the end and the sides of the petiole, is coincident with the rays of light. The pulvinus makes a sharp curvature which is concave to light, the blade of the leaflet being parallel to light. I have taken record of continuous action of strong light acting on the responding pulvinus of the leaflets from above. The result is an increasing positive curvature which reached a limit (Fig. 141). There was no neutralisation or reversal, demonstrating the absence of transverse conduction (*cf.* Fig. 132).

FIG. 141.—Positive para-heliotropic response of leaflets of *Erythrina indica*.

Para-heliotropic movement of leaflets of Mimosa pudica: *Experiment 143.*—These leaflets, as previously stated, fold themselves upwards, when strongly illuminated either from above or below. Diffuse electric stimulation also induce a closing movement upwards; hence the upper half of the pulvinule of these leaflets are the more excitable. In order to obtain a continuous record of the leaflet under the action of unilateral light, I constructed a very delicate recording lever magnifying about 150 times. Light of moderate intensity from a 100 candle-power incandescent lamp was applied on the less excitable lower side of the pulvinule. The record (Fig. 142) shows that the immediate response is positive, or a movement towards the light. But owing to transverse conduction, through the thin and highly conducting pulvinule, the response was quickly reversed into a very pronounced negative, or movement away from light. Had a delicate means of obtaining magnified record not been available, the slight positive twitch, and the gradual transition from positive to negative phototropic curvature would have passed unnoticed. Application of light from above gave, on account of the greater excitability of the upper half of the pulvinule, a pronounced positive response or movement towards light. The anomaly of an identical organ appearing as positively heliotropic when acted by light from above, and negatively heliotropic when acted from below, is now fully removed. The response of the leaflets is also seen to be determined by the directive action of light, though the short-lived response of the less excitable lower side is quickly masked by the predominant reaction of the more excitable upper side of the organ.

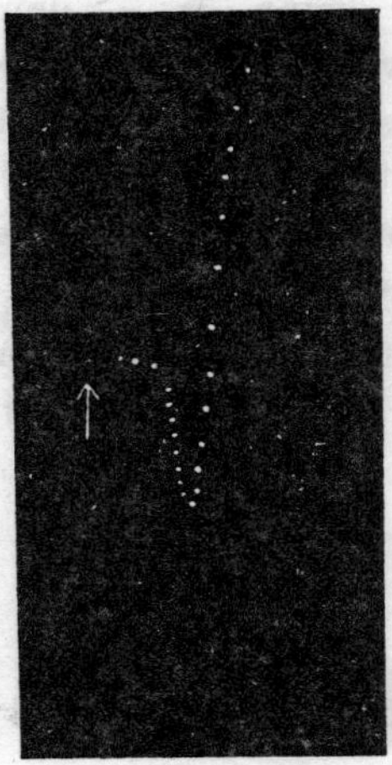

FIG. 142. FIG. 143.

FIG. 142.—Response of leaflet of Mimosa to light applied below: transient positive followed by pronounced negative curvature.

FIG. 143.—Response of leaflet of *Averrhoa*, to light applied above: transient positive followed by pronounced negative curvature.

Up-curve represents up-movement, and down-curve, down-movement.

NEGATIVE PARA-HELIOTROPISM.

Response of leaflet of Averrhoa carambola: *Experiment 144.*—The leaflets of this plant, and also those of *Biophytum sensitivum* fold downwards under action of strong light, applied above or below. In these leaflets diffuse electric stimulation induce a fall of the leaflets demonstrating the greater excitability of the lower half of the pulvinule. The analysis of reaction under light is rendered possible from the record of response of leaflet of *Averrhoa*, given in Fig. 143. Light of moderate intensity from an incandescent electric lamp acted from above: the result was a feeble and short-lived positive response, quickly reversed to strong negative by transmission of excitation to the more excitable lower side. Illumination from below gave rise only to strong positive response. Thus in *Averrhoa* the effect of continuous light applied above or below is a downward movement; in *Mimosa* the movement is upwards. The explanation of this difference lies in the fact, that in *Mimosa* leaflet it is the upper half of the pulvinule that is more excitable; while in *Averrhoa* and in *Biophytum* the lower is the more excitable half of the organ.

As a summary of the tropic action of light I shall give diagrammatic representations of various types of phototropic

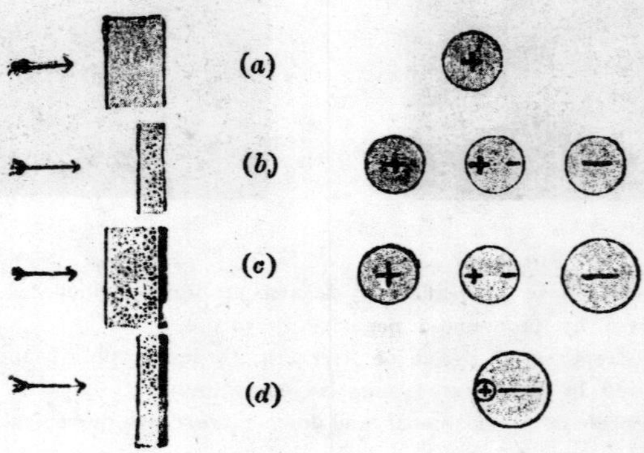

FIG. 144.—Diagrammatic representation of different types of phototropic response. (See text.)

response, including the photonastic (Fig. 144). The direction of the arrow indicates the direction of incident light. Dotted specimens are those which possess transverse conductivity. Thick lines represent the more excitable side of an anisotropic or dorsiventral organ. The size of the circles, with positive and negative signs, represents the amplitude and sign of curvature.

a. Radial thick organ, in which transverse conduction is absent. Curvature is *positive, i.e.*, movement towards light. The result will be similar when light strikes in an opposite direction, *i.e.*, from right to left.

b. Radial thin organ. There is here a possibility of transverse conduction. Sequence of curvature: *positive*, *neutral*, and *negative*. Reversal of direction of light gives rise to similar sequence of responses as before (*e.g.*, seedling of *Sinapis*).

c. Anisotropic thick organ; transverse conduction possible. Thick line represents the more excitable distal side. Sequence of curvature: positive, neutral and pronounced negative. When light strikes from opposite direction on the more excitable side the curvature will remain positive, since the pronounced reaction of the more excitable side cannot be neutralised or reversed by transmitted excitation to the less excitable distal side (*e.g.*, leaf of *Mimosa*).

In the absence of transverse conduction, the curvature remains positive (*e.g.*, leaflet of *Erythrina*).

d. Anisotropic thin organ with high transverse conductivity. Sequence of curvature: transient positive, quickly masked by predominant negative. Light striking on the more excitable side will give rise only to *positive*. The response in relation to the plant, will apparently be in the same direction whether light strikes the organ on one side or the opposite (*e.g.*, leaflets of *Mimosa*, *Averrhoa* and *Biophytum*).

I have shown that tissues in sub-tonic condition exhibit an acceleration of the rate of growth under stimulus (p. 224) the corresponding tropic reaction would therefore be away from stimulus or *negative* curvature. The tonic condition is, however, raised to the normal by the action of stimulus itself, and the tropic curvature becomes positive.

I give below a table which will show at a glance all possible variations of phototropic reaction.

TABLE XXXI.—MECHANICAL RESPONSE OF PULVINATED AND GROWING ORGANS UNDER LIGHT.

Description of tissue.	Action.	Effect observed.
I Tissue sub-tonic.	Stimulus causes increase of internal energy.	Expansion or enhanced rate of growth, *e.g.*, *Pileus* of *Coprinus* drooping in darkness, made re-turgid by light. Renewed growth of dark rigored plant exposed to light.
II Normally excitable organ under unilateral light. A. Organ radial.	A 1. Moderate light, causing excitatory contraction of proximal and positive expansion of distal.	1. Curvature towards light, *e.g.*, flower bud of *Crinum*.
	A 2. Strong light. Excitatory effect transmitted to distal, neutralising first.	2. Neutralisations, *e.g.*, seedling of *Setaria*.
	A 3. Intense and long-continued light. Fatigue of proximal and excitatory contraction of distal.	3. Reversed or negative response, *e.g.*, seedling of *Zea Mays*.
B. Dorsiventral organ.	B 1. Excitatory contraction of proximal predominant, owing either to greater excitability of proximal or feeble transverse conductivity of tissue.	1. Positive response, *e.g.*, upward folding of leaflets in so-called "diurnal sleep" of *Erythrina indica* and *Clitoria ternatea*.
	B 2. Transmission of excitation through highly conducting tissue to more excitable lower or distal. Greater contraction of distal.	2. Negative response, *e.g.*, downward folding of leaflets in so-called "diurnal sleep" of *Biophytum* and *Averrhoa*.
III Rhythmic tissue.	Considerable absorption of energy, immediate or prior.	Initiation of multiple response in *Desmodium gyrans* previously at standstill; multiple response under continuous action of light in *Biophytum*.

SUMMARY.

There is no line of demarcation between tropic and nastic movements.

In a differentially excitable organ the effect of strong unilateral stimulus becomes internally diffused, and causes greater contraction of the more excitable side of the organ.

In the absence of transverse conduction, the positive curvature reaches a maximum without neutralisation or reversal. The leaflets of *Erythrina indica* and of *Clitoria ternatea* thus fold upwards, the apices of the leaflets pointing towards the sun.

Internally diffused excitation under strong light induces greater contraction of the more excitable half of the pulvinule, causing upward folding of *Mimosa* leaflet, and downward folding of the leaflets of *Biophytum* and *Averrhoa*.

XXXV.—EFFECT OF TEMPERATURE ON PHOTOTROPIC CURVATURE

By

SIR J. C. BOSE,

Assisted by

GURUPRASANNA DAS.

I SHALL in this chapter deal with certain anomalies in phototropic curvature, brought about by variation of temperature and by seasonal change; certain organs again are apparently erratic in their phototropic response.

SEASONAL CHANGE OF PHOTOTROPIC ACTION.

Sachs observed a positive phototropic curvature in the stems of *Tropæolum majus* in autumn; but this was reversed into negative in summer; similarly in the hypocotyl of Ivy, the positive curvature in autumn is converted into negative curvature in summer.

Certain organs are apparently insensitive to the action of light. Thus no phototropic response is found in the tendril of *Passiflora* even under the action of strong light. The tendrils of *Vitis* and *Ampelopsis* exhibit, according to Wiesner, positive phototropism under feeble, and negative phototropism under strong light.

The anomalies referred to above may be explained by taking into consideration the modifying influence of temperature on the excitability, and the conductivity of the organ.

EFFECT OF TEMPERATURE ON EXCITABILITY.

The excitability of an organ is abolished at a low temperature; it is enhanced by a rise of temperature up to an optimum. The temperature minimum and optimum varies in different tissues. The following table shows the enhancement of excitability of *Mimosa* at different temperatures, the testing stimulus being the same.

TABLE XXXII—SHOWING VARIATION OF EXCITABILITY OF PULVINUS OF *Mimosa* AT DIFFERENT TEMPERATURES.

Temperature.	Amplitude of response.
22°C. ...	2 divisions.
27°C. ...	16 "
32°C. ...	36 "

Below 20°C. the excitability of the pulvinus of *Mimosa* is practically abolished. The excitability increases till an optimum temperature is reached, above which it undergoes a decline.

Though rise of temperature enhances excitability up to an optimum, there is an antagonistic reaction induced by it which opposes the excitatory contraction. The physiological reaction of a rise of temperature, within normal range, is expansion and this must oppose the contraction induced by stimulus. Hence the effect of rise of temperature is complex; it enhances the excitability which favours contraction, while tending to oppose this contraction by the induced physiological expansion. As a result of these opposite reactions there will be a critical temperature, below which the contractile effect will relatively be greater than expansion; above the critical point, expansion will be the predominant effect. The critical temperature will obviously be different in different organs. The positive curvature may thus be increased by a slight rise, while it may be neutralised, or even reversed by a greater rise of temperature.

The induced variation of excitability due to change of temperature is not the only factor in modifying tropic curvature, for variation of conductivity also exerts a marked effect.

EFFECT OF TEMPERATURE ON CONDUCTION.

The conducting power of an organ is greatly enhanced with rise of temperature. Thus in *Mimosa* the velocity of transmission of excitation is doubled by a rise of temperature through 9°C. (p. 100). An organ which is practically non-conducting at a low temperature will become conducting at a higher temperature.

Thus at a low temperature the organ may be non-conducting, and the excitatory contraction under unilateral stimulus will remain localised at the proximal side; this will give rise to a positive curvature. But under rising temperature, the power of transverse conduction will be increased and the excitation will be conducted to the distal side. The result of this will be a neutralisation or reversal into negative curvature (p. 139). A positive curvature is thus reversed into negative by change of excitability and conductivity, induced by rise of temperature; examples of this will be given presently.

PHOTOTROPIC RESPONSE OF TENDRILS.

I shall here adduce considerations which will show that the apparent anomalies regarding the response of tendrils to light is due to the variation of transverse conductivity of the organ. In a semi-conducting tissue, while the excitatory effect of feeble stimulus remains localised at the proximal side, the effect of stronger stimulus is conducted to the distal side. This explains the positive phototropic curvature of tendrils of *Vitis* and *Ampelopsis* under feeble light, and its reversal into negative curvature under intense light.

As the conducting power is increased with rise of temperature it is evident that at a certain temperature the tropic effect will be exactly neutralised by transverse conduction. Lowering of temperature, by reducing the transmission of excitation to the distal side, will restore the positive curvature. Enhancement of conduction under rise of temperature will, on the other hand, increase the antagonistic reaction of the distal side and give rise to a negative curvature.

I shall in verification of the above, describe experiments which I have carried out on the phototropic response of the tendril of *Passiflora*, supposed to be insensitive to the action of light.

Phototropic response of the tendril of Passiflora: *Experiment 145.*—The tendril was cooled by keeping it for a long time in a cold chamber, maintained at 15°C. The effect of unilateral light on the cooled specimen was found to be positive; the tendril was next allowed to assume the temperature of the room which was 30°C. The response was now found to have undergone a change into negative. The positive and negative phototropic curvatures of an identical organ at different temperatures is seen in the two records given in figure 145. Neutralisation takes place at an intermediate temperature, and the organ thus appears insensitive to light.

SEASONAL VARIATION OF PHOTOTROPIC CURVATURE.

Reference has been made of the phototropic curvature of *Tropæolum* and of Ivy undergoing a change from

FIG. 145.—(*a*) Positive curvature of tendril of *Passiflora* at 15°C.; (*b*) negative phototropic curvature at 30°C.

positive in autumn to negative in summer. The experiment described above shows that rise of temperature, by enhancing transverse conductivity, transforms the positive into negative heliotropic curvature. The reversal of the phototropic curvature of *Tropæolum* and Ivy, from positive in autumn to negative in summer, finds a probable explanation in the higher temperature condition of the latter season. This inference finds independent support from the fact previously described (p. 100) that while the velocity of conduction of excitation in the petiole of *Mimosa* is as high as 30 mm. per second in summer, it is reduced to about 4 mm. in late autumn and early winter.

ANTAGONISTIC EFFECTS OF LIGHT AND OF RISE OF TEMPERATURE.

I have explained the complex effect of rise of temperature on phototropic curvature. Rise of temperature, within limits, enhances the excitability, and therefore the positive curvature under light. Its expansive reaction, on the other hand, opposes the contraction of the proximal side, which produces the normal positive curvature. Rise of temperature, as previously stated, introduces another element of variation by its effect on conductivity. Transverse conduction favoured by rise of temperature promotes neutralisation and reversal; the resultant effect will thus be very complicated. I give below account of an experiment where the induced positive curvature under light underwent a reversal during rise of temperature.

Reversal of tropic curvature under rise of temperature: Experiment 146.—The specimen employed for this experiment was a seedling of pea, enclosed in a glass chamber, the temperature of which could be gradually raised by means of an electric heater. Provisions were made to maintain the chamber in a humid condition. The temperature of the chamber was originally at 29°C., and application of light on one side of the organ gave rise to positive curvature, followed by complete recovery on the cessation of light (Fig. 146a). The next experiment was carried out with the same specimen; while the plant was undergoing increasing positive curvature under the continued action of light, the temperature of the chamber was gradually raised from 29° to 33°C. at the point marked with arrow. It will be seen that the positive curvature became arrested, neutralised, and finally reversed into negative (Fig. 146b).

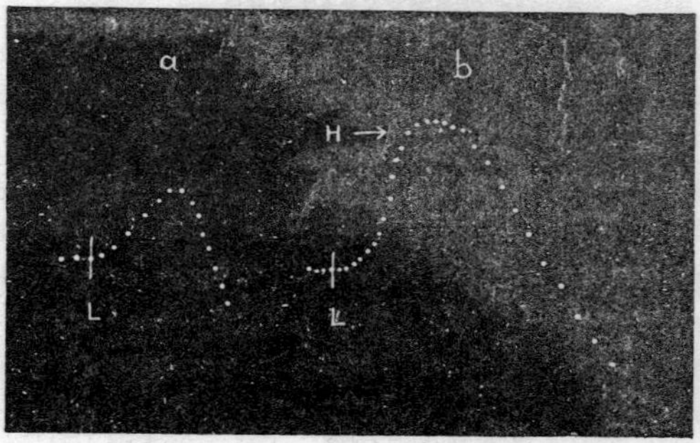

FIG. 146.—Effect of rise of temperature on phototropic curvature. (*a*) normal positive curvature followed by recovery, (*b*) reversal of positive into negative curvature by rise of temperature at (H). (Pea seedling).

After-effect of rise of temperature: Experiment 147.—The after-effect of rise of temperature exhibited by this specimen was extremely curious. The temperature of the chamber was allowed to return to the normal, and the experiment repeated after an hour; the response was now found to be negative (Fig. 147a). It appeared probable that the temperature in the interior of the tissue had not yet returned to the normal, and an interval of four hours was therefore allowed for the restoration of the tissue to the normal temperature of the room. The response still persisted to be negative, as seen in the series of records obtained under successive stimulations of light of short duration; these negative responses exhibited recovery on the cessation of light (Fig. 147b). This reversal of response as an after-effect of rise of temperature was in this case found to

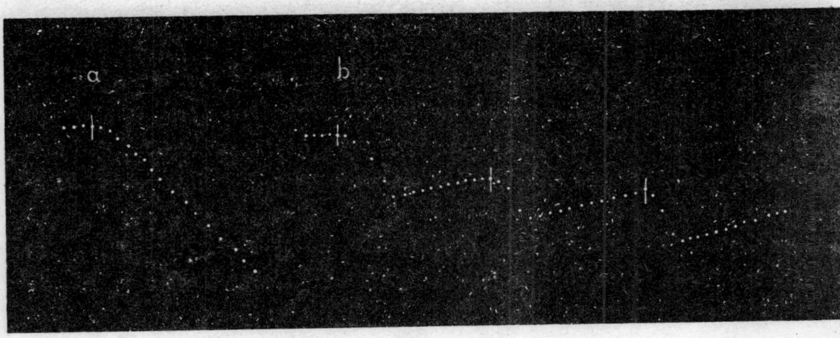

FIG. 147.—After-effect of rise of temperature, persistent negative curvature: (*a*) response one hour after rise of temperature; (*b*) series of negative responses after 4 hours (successive stimuli applied at vertical lines).

persist for several hours. I experimented with the same specimen next day when the response was found restored to the normal positive.

SUMMARY.

Rise of temperature, within limits, enhances the general excitability of the organ. This has the effect of increasing positive phototropic curvature. But the physiological expansion induced by rise of temperature exerts an antagonistic effect.

The transverse conductivity is increased with the rise of temperature ; this favours neutralisation and reversal of phototropic curvature.

Tendrils of *Passiflora*, supposed to be phototropically insensitive, exhibit positive curvature at low, and negative curvature at a moderately high temperature.

The change of phototropic curvature exhibited by *Tropæolum majus* and Ivy, from positive in autumn to negative in summer, is probably due to the effect of temperature. Higher temperature with enhanced transverse conductivity in summer, may thus convert positive into negative curvature.

The physiological effects of rise of temperature and the stimulus of light are antagonistic to each other.

Rise of temperature tends to neutralise or reverse the positive phototropic curvature. The after-effect of temperature is often very persistent.

XXXVI.—ON PHOTOTROPIC TORSION

By

SIR J. C. BOSE,

Assisted by

SURENDRA CHANDRA DAS.

IN addition to positive or negative curvatures light induces a responsive torsion. With regard to this Jost says:—

"The mechanics of the torsions have not as yet been fully explained. It has long been believed that these torsions were occasioned only by the action of a series of external factors, such as light, gravity, weight of the organ which individually led to curvatures, but in combination induced torsions; but later investigations have shown that torsions might appear when light only was the functional external factor. . . . If the torsions cannot generally be regarded as due to the combination of two curvatures, we are completely in the dark as to the mechanics of their production."*

A leaf when struck laterally by light undergoes a twist, so that the upper surface is placed, more or less, at right angles to the incident rays; as no explanation was available for this movement, the suggestion has been made that the particular reaction is for the advantage of the plant. I shall presently show, that it is possible to reverse this normal torsion and thus make the upper surface of the leaf move away from light.

* Jost—*Ibid*—p. 465.

The experiments which I shall presently describe will, it is hoped, throw light on the obscure phenomenon. I shall be able to show :

(1) that the torsional response is not dependent on the combination of two curvatures,

(2) that it is also independent of the effect of weight,

(3) that it may be induced not merely by stimulus of light but by all forms of stimulation,

(4) that the direction of the torsional response depends on the direction of the incident stimulus and the differential excitability of the organ, and

(5) that there is a definite law which determines the torsional movement.

EXPERIMENTAL ARRANGEMENTS.

I shall first describe a typical experiment on the responsive torsion under the action of light. We have seen that in the pulvinus of *Mimosa*, light of moderate intensity and of short duration applied on the upper half induces a slow up-movement, while the stimulus of light applied below induces a more rapid down-movement. The difference is due to the fact that the lower half of the pulvinus is relatively the more excitable. Vertical light thus induces a movement in a vertical plane. But an interesting variation is induced in the response under the action of lateral light. A stimulus will be called *lateral* when it acts on either the right or left flank of a *dorsiventral* organ. We shall presently find that a dorsiventral organ responds to lateral stimulus by torsion.

The present series of experiments were carried out with the leaf of *Mimosa*, and in order to eliminate the effect of weight and also for obtaining record of pure torsion, I employed the following device. The petiole was enclosed in a hooked support made of thin rod of glass, the petiole resting on the concavity of the smooth surface. Friction and the effect of weight is thus practically eliminated; the looped support prevented up or down movements, and yet allowed perfect freedom for torsional response. This latter is magnified by a piece of stout aluminium wire fixed at right angles to the

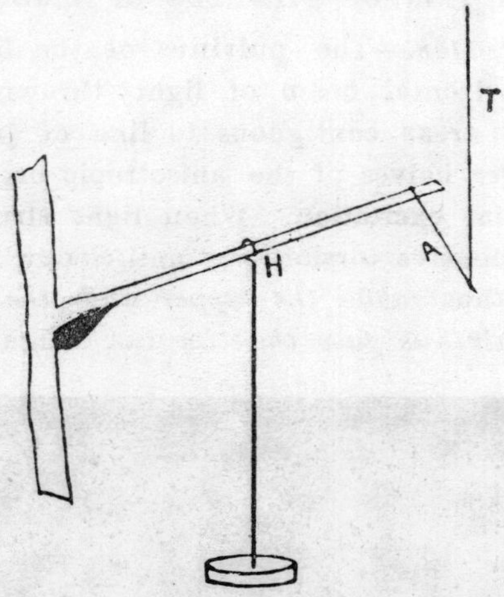

FIG. 148.—Diagrammatic representation for record of torsional response. H, thin glass hook: A, aluminum wire attached to petiole for magnification of torsional movement. T, silk thread for attachment to recording lever.

petiole (Fig. 148). The end of the aluminium wire is attached to the short arm of a recording lever; there is thus a compound magnification of the torsional movement. The Oscillating Recorder gave successive dots at intervals which could be varied from 20 seconds to 2 minutes. Time-relations of the response may thus be obtained from the dotted record.

With the experimental device just described, we shall be in a position to study the effect of various stimuli applied at one flank of the pulvinus—at the junction of the upper and lower halves of the organ. The observer standing in front of the leaf is supposed to look at the stem. Torsional response will then appear as a movement either with or against the hands of the clock. The torsional response, right-handed or left-handed, will presently be shown to depend on the direction of incident stimulus. In figure 149, anti-clockwise torsion is recorded as an up-curve; clockwise rotation is recorded as a down-curve.

ACTION OF STIMULUS OF LIGHT.

Experiment 148.—The pulvinus of the leaf was stimulated by a horizontal beam of light thrown in a lateral direction; the areas contiguous to line of junction of the upper and lower halves of the anisotropic organ thus underwent differential excitation. When light struck on the left flank, the responsive torsion was anti-clockwise; the responsive reaction thus made *the upper and the less excitable half of the pulvinus face the stimulus.* Figure 149 gives a

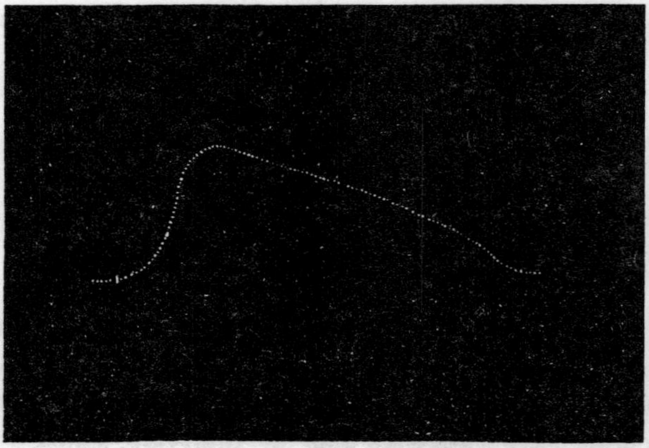

FIG. 149.—Record of torsional response of pulvinus of *Mimosa pudica.*

record of the torsional movement when light struck the left flank of the organ; on the cessation of stimulus the response is followed by recovery.

DIRECTIVE ACTION OF STIMULUS.

Experiment 149.—If now the direction of stimulus be changed so that light strikes on the right flank instead of the left, the responsive torsion is found to be reversed, the direction of movement being clockwise. Here also the responsive movement is such that it is the less excitable upper half of the organ that is made to face the stimulus. It will thus be seen that the torsion, anti-clockwise or clockwise, depends on two factors, namely the direction of stimulus, and the differential excitability of the organ.

EFFECT OF DIFFERENT MODES OF LATERAL STIMULATION.

I shall now proceed to show that the torsional response is induced not merely by the action of light, but by all forms of stimulation.

Effect of chemical stimulation: Experiment 150.—Dilute hydrochloric acid was at first applied on the left flank of the pulvinus along the narrow strip of junction of the upper and lower halves. This gave rise to a responsive torsion against the hands of a clock. Chemical stimulation of the right flank induced, on the other hand, a torsional movement with the hands of a clock. Here also the direction of stimulus is found to determine the direction of responsive torsion.

Effect of thermal radiation: Experiment 151.—I next employed thermal radiation as the stimulus; the source of radiation was a length of electrically heated platinum wire. It is advisable to interpose a narrow horizontal slit, so as to localise the stimulus at the junction of the upper and lower halves of the pulvinus. Stimulus applied at the left flank induced left-handed or anti-clockwise torsion; application at the right flank gave rise to right-handed torsion.

Geotropic stimulus.—The stimulus of gravity induces, as I shall show in a subsequent chapter, a similar responsive torsion, the direction of which is determined by the direction of the incident stimulus.

EFFECT OF DIFFERENTIAL EXCITABILITY ON THE DIRECTION OF TORSION.

Under normal conditions, the torsional response under light places the upper surface of the leaf or leaflets at right angles to light. That this movement is not due to some specific sensibility to light is shown by the fact that all modes of stimulation, chemical, thermal or gravitational, induce similar responsive torsion. The torsional response is, moreover, shown to be determined by the direction of incident stimulus, and the differential excitability of the organ. This latter may be reversed by the local application of various depressing agents on the normally more excitable lower half of the pulvinus. Under this treatment, the lower half of the pulvinus may be rendered relatively the less excitable. Lateral application of light now induces a torsional movement which is the reverse of the normal, so that the upper surface of the leaf moves away from light. The advantage of the plant cannot, therefore, be the factor which determines the directive movement; the teleological argument often advanced is, in any case, no real explanation of the phenomenon.

In all the instances given above, and under every mode of stimulation, the responsive movement makes the less excitable half of the pulvinus face the stimulus. The torsional response is, in reality, the mechanical result of the differential contraction of a complex organ, which is fixed at one end and subjected to lateral stimulation. I have been able to verify this, by the construction of an artificial pulvinus consisting of a compound strip, the upper half of which is ebonite, and lower half the more contractile stretched India-rubber; if such a strip be held securely at one end in a clamp, and if the lateral flank, consisting half of ebonite and half of India-rubber, be subjected to radiation, and record taken in the usual manner, it will be found that a torsional response takes place which is similar to that of the pulvinus of *Mimosa*. The above experiment was devised to offer an explanation of the mechanics of the movement. It should, however, be borne in mind in this connection that the torsional response of pulvinus is brought about by differential *physiological* contraction of the organ, the movement being abolished at death.

From the results given above, we arrive at the following:—

LAWS OF TORSIONAL RESPONSE.

1. AN ANISOTROPIC ORGAN, WHEN LATERALLY EXCITED BY ANY STIMULUS, UNDERGOES TORSION BY WHICH THE LESS EXCITABLE SIDE IS MADE TO FACE THE STIMULUS.

2. THE INTENSITY OF TORSIONAL RESPONSE INCREASES WITH THE DIFFERENTIAL EXCITABILITY; WHEN THE ORIGINAL DIFFERENCE IS REDUCED, OR REVERSED, THE TORSIONAL RESPONSE UNDERGOES CONCOMITANT DIMINUTION OR REVERSAL.

Having thus established the laws that guide torsional response, I shall try to explain certain related phenomena which are regarded as highly obscure. I shall also describe the application of the method of torsional response in various investigations.

COMPLEX TORSION UNDER LIGHT.

The leaves of the so-called "Compass plants" exhibit very complex movements, these being modified according to the intensity of incident light. Thus in compass plants the leaves, under moderate intensity of light in the morning or in the evening, turn themselves so as to expose their surfaces to the incident rays. But under intense sun light, the leaves perform bendings and twistings so that they stand at profile at midday.

I have not yet been able to secure "Compass plants" at Calcutta. I shall, however, describe my investigations on the complicated torsional movements exhibited by certain leaflets by the action of vertical light. The results obtained from these will show that torsional movements, even the most complex, are capable of explanation from the general laws that have been established.

Torsional movement of leaflet of Cassia alata : *Experiment 152.*—These leaflets are closed laterally at night but place themselves in an outspread position at day time. The character of the movement is, however, modified by the intensity of light. With moderate light in the morning the leaflets open out laterally. But under more intense light, the pulvinules of the leaflets exhibit a torsion by which the formerly infolded surfaces of the leaflets are

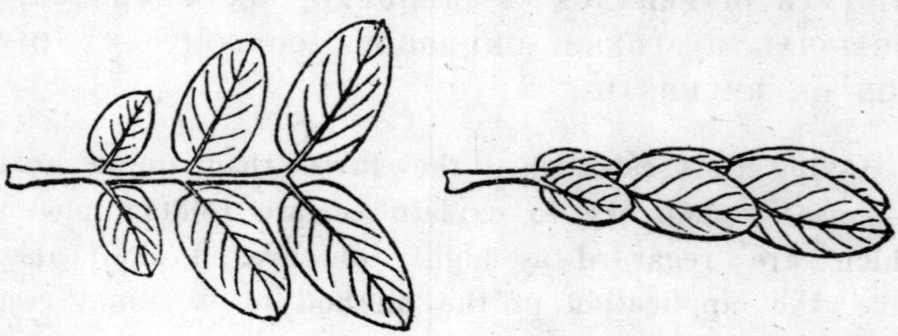

FIG. 150.—Leaflets of *Cassia alata* : open in daytime, and closed in evening.

exposed at right angles to light from above (Fig. 150). Such complicated movements, in two directions of space, are also exhibited by other leaflets which are closed at night in a lateral direction.

For obtaining an explanation of these complex movements under different intensities of light, we have first to discover the particular disposition of the two halves of the pulvinule which are unequally excitable; we have next to explain the responsive movements under the directive action of moderate and of intense light.

Determination of differential excitabilities of the organ: Experiment 153.—In the leaflet of *Cassia* the movement of opening under diffuse stimulation of light can only be brought about by the contraction of the outer half, which must therefore be the more excitable. This is independently demonstrated by the reaction to an electric-shock. On subjecting the half closed leaflets to diffuse electric stimulation, they open outwards in a *lateral* direction. The disposition of the unequally excitable halves of the pulvinule is thus different from that of the main pulvinus of *Mimosa*. In the latter, the plane that divides the two halves is horizontal, the lower half being the more excitable. Thus in the pulvinule of *Cassia* the plane that separates the two unequally excitable halves is vertical, the outer half being the more excitable than the inner. By inner half is here meant that half which is inside when the leaflets are closed.

Effect of strong vertical light: Experiment 154.—When the plant is placed in a moderately lighted room, the leaflets open out laterally to the outmost. This is brought about by the contraction of the more excitable outer half of the organ. If strong light be thrown down from above, a new movement is superposed, namely, of torsion by which the leaflets undergo a twist and thus place their inner surface at right angles to the vertical light. In order to investigate this phenomenon in greater detail I placed the plant in a well lighted room, the leaflets being three quarters open under the diffuse light. A very light index was attached to the leaflet for magnifying the subsequent torsional movement. A strong beam of parallel light from an arc lamp was thrown down on the pulvinule from above ; this fell at the junction of the more excitable outer with the less excitable inner half of the organ, the plane of separation of the two unequally excitable halves being, as previously explained, vertical. I have shown that under lateral stimulation, a differentially excitable organ undergoes torsion by which the less excitable half is made to face the stimulus. Since it is the inner half of the organ that is the less excitable, the attached leaflet becomes twisted so as to expose its (former infolded) surface upwards, at right angles to the incident light.

As a confirmatory test, strong light was made to strike the pulvinule from *below* with the result that the leaflets exhibited an opposite torsion by which their surfaces faced downwards, so as to be at right angles to light that struck them from below.

Under normal conditions sunlight comes from above ; stimulation thus takes place at the junction of the two differentially excitable halves of the organ, the plane of separation of which is vertical. The torsion induced makes the less excitable inner half turn in such a way that the inner surfaces of the leaflets are placed perpendicular to the incident light.

ADVANTAGES OF THE METHOD OF TORSIONAL RESPONSE.

The torsional response not only affords a new method of enquiry on the reaction of various stimuli, but it also possesses certain advantages. For instance in studying the response of the leaf of *Mimosa* under light, the records were taken of the movement of the leaf in a vertical plane. But the responsive up-movement, induced by light acting from above, is opposed by the weight of the leaf. But in the torsional response, the leaf rests on the hooked glass support and the movement is thus free from the complicating factor of the weight of the leaf. Again the pulvinus of *Mimosa*, for example, is sometimes subject to spontaneous variation of turgor, on account of which it exhibits an autonomous up or down movement. In the ordinary method of record the true response to external stimulus may thus be modified by natural movement of the leaf. But in the torsional method, the autonomous up or down movement is restrained by the hooked support, and the response to lateral stimulus is unaffected by the spontaneous movement of the leaf. The torsional method, moreover, opens out possibilities of inquiry in new directions, such as the comparison of the excitatory effects of different stimuli by the Method of of Balance, and the determination of the effective direction of geotropic stimulus.

THE TORSIONAL BALANCE.

A beam of light falling on the left flank of the pulvinus of *Mimosa* induces a torsion against the hands of the clock. A second beam falling on the right flank opposes the first movement; the resultant effect is therefore determined by the effective stimulation of the two flanks. The pulvinus thus becomes a delicate index by which two stimuli may be compared with each other. The following experiment is cited as an example of the application of the method of phototropic balance.

Experiment 155.—Parallel beam of light from a small arc lamp passing through blue glass falls on the left flank of the pulvinus; a beam of blue light also strikes the pulvinus from the right side, and the intensity of the latter is so adjusted that the resultant torsion is zero. Blue glass is now removed from the left side, the unobstructed white light being allowed to fall on the left flank of the pulvinus. This was found to upset the balance, the resultant torsion being anti-clockwise. This showed that white light induced greater excitation than blue light. We next interpose a red glass on the left side, with the result that the balance is upset in the opposite direction. This is because the phototropic effect of red light is comparatively feeble. We may thus compare the tropic effect of one form of stimulus against a totally different form, phototropic against geotropic action for example. It is enough here to draw attention to the various investigations rendered possible by the method of balance. Concrete examples of some of these will be given in a subsequent chapter.

DETERMINATION OF THE DIRECTION OF STIMULUS.

I have shown that the torsion, clockwise or anti-clockwise, is determined by the direction of incident stimulus. Hence it would be possible to determine the direction of incident stimulus from the observed torsional movement. In the case of light, the direction of incident stimulus is quite apparent. But it is difficult to determine the direction of stimulus which is itself invisible. In such cases, the torsional movement gives us infallible indication of the effective direction of stimulus. The application of this principle will be found in a later chapter.

SUMMARY.

Lateral stimulus induces a torsional response in a dorsiventral organ. This is true of all modes of stimulation.

The responsive torsion is determined by the direction of incident stimulus, and the differential excitability of two halves of the organ, the torsion being such that the less excitable half of the organ is made to face the stimulus.

The twist exhibited by various leaves and leaflets under light finds its explanation from the demonstrated laws of torsional response.

The direction of incident stimulus may be determined from the responsive torsion of a dorsiventral organ.

The Method of Torsional Balance enables us to compare the excitatory efficiencies of two different stimuli which act simultaneously on the two flanks of the organ.

XXXVII.—RADIO-THERMOTROPISM

By

SIR J. C. BOSE,

Assisted by

GURUPRASANNA DAS.

WE have studied the tropic curvature induced by different rays of light. We saw that while the more refrangible rays of the spectrum were most effective, the less refrangible rays were ineffective. Below the red, there are the thermal rays about whose tropic effect very little is definitely known.

The intricacies of the problem are very great owing to the difficulty of discriminating the effect of temperature from that of radiation; to this must be ascribed the contradictory results that have been obtained by different observers, of which Pfeffer gives the following summary:*

"In addition to the action of ultra-red rays which are associated with the visible part of the spectrum, dark heat-rays of still greater wave length, as well as differences of temperature may produce a thermotropic curvature in certain cases. Wortmann observed that seedlings of *Lepidium sativum* and *Zea Mays*, as well as sporangiphores of *Phycomyces* curved towards a hot iron plate emitting dark heat-rays. Steyer has, however, shown that the sporangiphore of *Phycomyces* has no power of thermotropic reaction . . Wortmann observed that the seedling shoot of

* Pfeffer—*Ibid*—Vol. III, p. 776.

Zea Mays was positively, but that of *Lepidium* negatively, thermotropic . . . Steyer, however, found that both plants were positively thermotropic. Wortmann has also investigated the radicles of seedlings by growing them in boxes of saw-dust, one side being kept hot, the other cold."

It will be noted that in the investigations described above, thermotropic reaction has been assumed to be the same under variation of temperature (as in experiments with unequally heated saw-dust), and under radiation from heated plate of metal. With reference to this Jost maintains that "so far as we know, thermotropism due to *radiant* heat cannot be distinguished from thermotropism due to *conduction*."*

The effect of temperature, within optimum limits, is a physiological expansion and enhancement of the rate of growth. The effect of visible radiation is, on the other hand, a contraction and retardation of growth. Should radiant heat act like light, the various tropic effects in the two cases would be similar; the temperature effect would in that case be opposite to the radiation effect. In order to find whether the thermal radiation produces tropic curvature similar to that of light, we have to devise a crucial experiment in which the complicating factor of rise of temperature on the responding organ is eliminated.

Experiment 156.—I have described the effect of light applied unilaterally to the stem of *Mimosa*, at a point diametrically opposite to the indicating leaf (*Expt.* 104). It was shown that the effect of indirect stimulus induced at first an erectile movement of the leaf, and that this was followed by a fall of the leaf on account of transverse transmission of excitation. In the present experiment I applied thermal radiation instead of light. The source of radiation was a spiral of platinum wire heated short of incandescence by

* Jost—*Ibid*—p. 480.

means of electric current. The intensity of incident radiation could thus be maintained constant, or increased or decreased by approach or recession of the radiating spiral. The effect of unilateral stimulus of heat-rays was found exactly similar to that of light; *i.e.*, there was at first an erectile movement due to indirect stimulation, followed by the fall of the leaf due to transmitted excitation. It will be noticed that under the particular condition of the experiment, the responding pulvinus was completely shielded from temperature-variation. The reaction to thermal radiation is thus similar to that of light.

As regards the effects of rise of temperature and radiation I have shown that they are antagonistic to each other (pp. 211, 308). Thus in positive types of thermonastic organs like the flower of *Zephyranthes*, while rise of temperature induces a movement of opening, radiation causes the opposite movement of closure. Again, in the negative type exemplified by *Nymphæa*, rise of temperature induces a movement of closure; radiation on the other hand, brings about the opposite movement of opening. The tropic effect of thermal radiation thus takes place in opposition to that of rise of temperature, and the resultant effect is therefore liable to undergo some modification, depending on the relative sensibility of the organ to radiation and to variation of temperature.

The facts that have been given above prove that infrared radiation is as effective a mode of stimulation as the more refrangible rays of the spectrum. Phototropic and radio-thermotropic reactions would therefore prove to be essentially similar. The following experiments fully confirm the similarity of the two reactions.

POSITIVE RADIO-THERMOTROPISM.

Experiment 157.—I shall now describe the normal reaction of a growing organ to the unilateral stimulus of thermal radiation. Figure 151 gives a record of response of the stem of *Dregea* to stimulus of short duration ; the

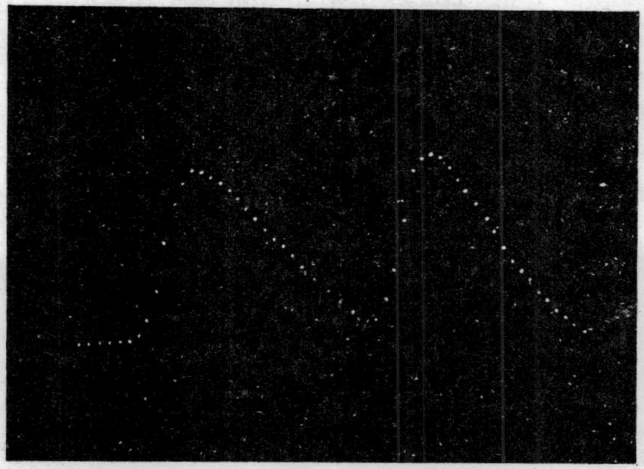

FIG. 151.—Positive response to short exposure to thermal radiation. Successive dots at intervals of 5 seconds. (*Dregea volubilis.*)

induced curvature is positive or towards the source of heat. On the cessation of stimulus, there is a recovery which is practically complete, and which takes place at a slower rate than the excitatory positive curvature. Repetition of stimulus gives rise to responses similar to the first. *Successive stimuli of moderate intensity thus give rise to repeated responses of growth curvature.* An arbitrary distinction has been made between the responses of pulvinated and of growing organs. The former is distinguished as a movement of variation, with its supposed characteristic of repeated response. But the experiment described shows that this is also met with in the response by growth curvature. It is only under long continued stimulation that the curvature is fixed by growth.

DIA-RADIO-THERMOTROPISM.

The positive curvature is induced by retardation of growth at the proximal side, and enhancement of growth at the distal side. This latter effect is, as we have seen, brought about by the effect of indirect stimulation.

But under long continued action of stimulus, the negative or excitatory impulse reaches the distal side, inducing diminution of turgor and retardation of the rate of growth. This leads to neutralisation, the organ placing itself at right angles to the orienting stimulus.

Experiment 158.—This neutralisation is seen in the

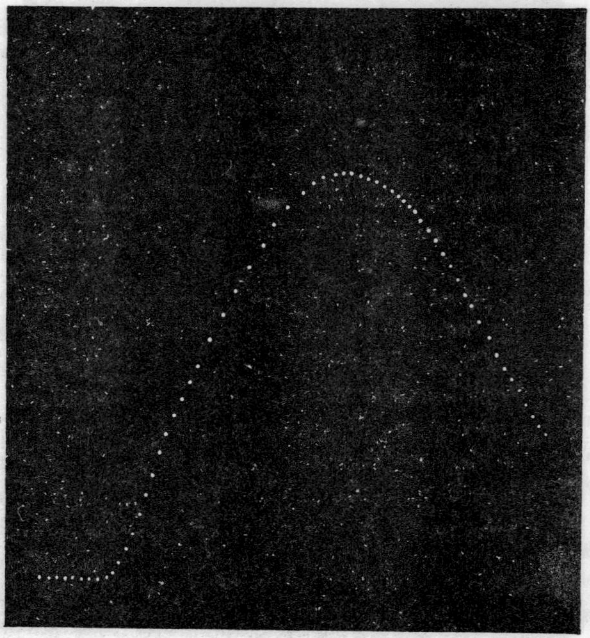

FIG. 152.—Record of positive, neutral and reversed negative curvature under continued action of thermal radiation. The negative response went off the plate. Successive dots at intervals of 5 seconds. (*Dregea volubilis*).

record given in figure 152, where under continuous unilateral stimulation, the growing organ exhibited its maximum positive curvature, after which the movement became arrested by the arrival of the excitatory impulse at the distal side, on account of which the first positive curvature became neutralised. Further continuation of stimulus caused a reversal into negative in the course of 7 minutes. It will thus be seen that in inducing phototropic curvature, the heat rays in sunlight play as important a part as the more refrangible rays of the spectrum.

SUMMARY.

The effects of rise of temperature and of radiation are antagonistic to each other.

Under unilateral action of thermal radiation a positive curvature is induced by the retardation of growth at the proximal, and acceleration of growth at the distal side of the organ.

There is a complete recovery on the cessation of stimulus of moderate intensity and short duration. Repeated responses may thus be obtained similar to repeated responses in pulvinated organs. In certain tissues the power of conduction in a transverse direction is wanting; excitation remains localised at the proximal side, and the responsive curvature remains positive.

In other cases, there is a slow conduction of excitation to the distal side. The result of this under different circumstances is dia-radio-thermotropic neutralization, or a reversed negative curvature.

In inducing phototropic curvature, the heat rays in sunlight play as important a part as the more refrangible rays of the spectrum.

XXXVIII.—RESPONSE OF PLANTS TO WIRELESS STIMULATION

By

SIR J. C. BOSE,

Assisted by

GURUPRASANNA DAS.

A GROWING plant bends towards light, and this is true not only of the main stem but also of its branches and attached leaves and leaflets. Light affects growth, the effect being modified by the intensity of radiation. Strong stimulus of light causes a diminution of the rate of growth, but very feeble stimulus induces an acceleration. The tropic effect is very strong in the ultra-violet region of the spectrum with its extremely short wave length, but the effect declines practically to zero as we move towards the less refrangible rays—the yellow and the red with their comparatively long wave length. As we proceed beyond the infra-red region, we come across the vast range of electric radiation, the wave lengths of which vary from 0·6 cm., the shortest wave I have been able to produce, to others which may be miles in length. There thus arises the very interesting question, whether plants perceive and respond to the long ether waves including those employed in signalling through space.

At first sight this would appear to be very unlikely; for the most effective rays are in the ultra-violet region with wave length as short as 20×10^{-6} cm.; but with electric waves used in wireless signalling we have to deal with waves 50 million times as long. The perceptive power of our retina is confined within the very narrow range of a single octave, the wave lengths of which lie between 70×10^{-6} cm. and 35×10^{-6} cm. It is difficult to imagine that plants could perceive radiations so widely separated from each other as the visible light and the invisible electric radiation.

But the subject assumes a different aspect, when we take into consideration the *total* effect of radiation on the plant. Light induces two different effects which may broadly be distinguished as external and internal. The former gives rise to movement; the latter finds no outward manifestation, but consists of an 'up' or assimilatory chemical change, with concomitant increase of potential energy. Of the two reactions then, one is dynamic attended by dissimilatory 'down' change; the other is potential, associated with the opposite 'up' change. In reality the two effects take place simultaneously; but one of these becomes predominant under definite conditions.

The modifying condition is the *quality* of light; with reference to this I quote the following from Pfeffer: "So far as is at present known, the action of different rays of the spectrum gives similar curves in regard to heliotropic and phototactic movements, to protoplasmic streaming and movements of the chloroplastids as well as the photonastic movements produced by growth or by changes of turgor. On the other hand, it is the less refrangible rays which are most active in photo-synthesis."* The dynamic and potential manifestations are thus seen to be complementary to each other, the rays which induce photo-synthesis being relatively ineffective for tropic reaction and *vice versâ*.

Returning to the action of electric waves, since they exert no photo-synthetic action they might conceivably induce the complementary tropic effect. These considerations led me to the investigation of the subject fourteen

* Pfeffer—Vol. II, p. 104.

years ago, and my results showed that very short electric waves induce a retardation of rate of growth; they also produce responsive movements of the leaf of *Mimosa*, when the plant was in a highly sensitive condition.* The energy of the short electric waves is very feeble, and undergoes great diminution at a distance; hence the necessity of employment of a specimen of plant in a highly sensitive condition.

I resumed my investigations on the subject at the begining of this year. I wished to find out whether plants in general perceived and responded to the long ether waves which reached it from a distance. The perception of the wireless stimulation was to be tested not merely by the responsive movement of sensitive plants, but by diverse modes of response given by all kinds of plants.

Stimulus induces, as we have seen, three different types of response in plants. It causes excitation in sensitive plants like *Mimosa*, in consequence of which the leaf undergoes a fall; this is the mechanical response to stimulus. Stimulus also induces electric response in plants, both sensitive and ordinary, the excited tissue undergoing an electric change of galvanometric negativity. Finally, the effect of stimulus on growing plants is a variation in the rate of growth, an acceleration under feeble, and a retardation under strong stimulus. I undertook to investigate the effect of electric waves on plants by the methods of mechanical and of electrical responses, and also by that of induced variation of growth.

THE WIRELESS SYSTEM.

For sending wireless signals, I had to improvise the following arrangement, more powerful means not being

* "Plant Response" —p. 618 (1905).

available. The secondary terminals of a moderate sized Ruhmkorff's coil were connected with two cylinders of brass, each 20 cm. in length; the sparking took place between two small spheres of steel attached to the cylinders. One of the two cylinders was earthed, and the other connected with the aerial 10 meters in height. The receiving aerial was also 10 meters in height and its

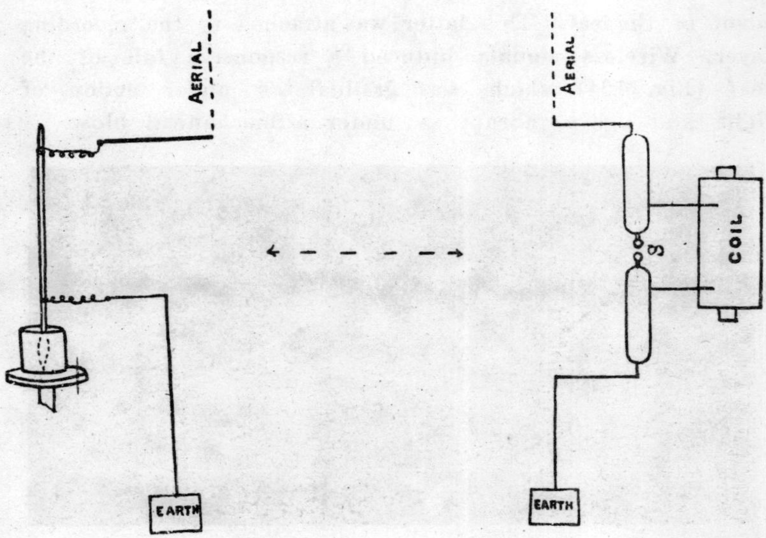

FIG. 153.—Diagrammatic representation of method employed for obtaining response to wireless stimulation. Transmitting apparatus seen to the right. Receiving aerial connected to upper part of plant, the lower part of the plant or the flower-pot being connected with the earth.

lower terminal led to the laboratory, and connected by means of a thin wire to the experimental plant growing in a pot; this latter was put in electric connection with the earth (Fig. 153). The distance between the transmitting and receiving aerial was about 200 meters, the maximum length permitted by the grounds of the Institute.

MECHANICAL RESPONSE OF *Mimosa.*

Experiment 159.—One of the leaves of *Mimosa* was connected with the aerial by means of a thin tinsel of loose wire, which did not interfere with the free movement of the leaf. This latter was attached to the recording lever. Wireless signals induced a responsive fall of the leaf (Fig. 154) which was gradual as under action of light, and not so abrupt as under a mechanical blow.

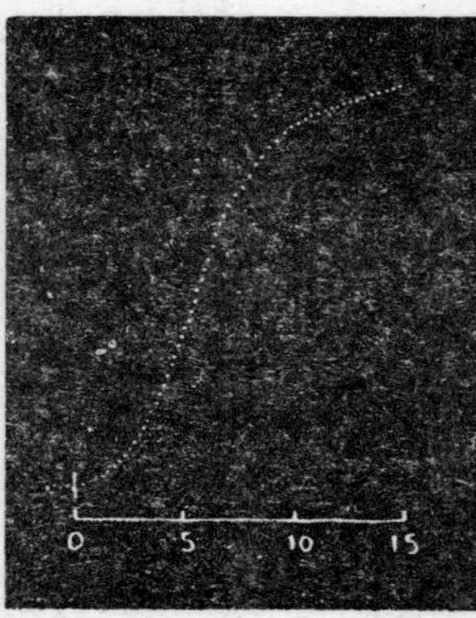

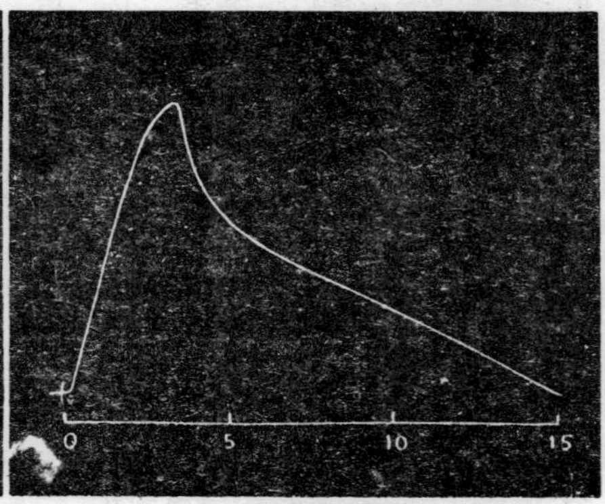

FIG. 154. FIG. 155.

FIG. 154.—Mechanical response of leaf of *Mimosa* to electric wave.

FIG. 155.—Electric response of *Mimosa pudica* to wireless stimulation.

THE ELECTRIC RESPONSE.

Experiment 160.—The leaf of *Mimosa* was in this experiment held securely, and two electrical connections made, one with the less excitable upper and the other with the more excitable lower half of the pulvinus. The incident ether-wave induced an electric response in the pulvinus, the more excitable lower half exhibiting galvanometric negativity. On the cessation of stimulus there was a recovery (Fig. 155).

It is not at all necessary to employ the sensitive *Mimosa* for exhibition of electric response; for this is universally exhibited by all plants. The only condition for electric response is that the points of electric contacts should be made with two unequally excitable areas in the plant. This may be secured by artificial means as by causing 'injury' to one point of contact.* It is however much better to take advantage of the natural difference of excitability of two different areas in the organ as in the pulvinus of *Mimosa*. This difference of excitability is also found between the inner and outer sides of a hollow tubular organ as in the peduncles of various lilies. I was thus able to secure specimens which were far more sensitive to the action of electric waves than the pulvinus of *Mimosa*.

EFFECT OF WIRELESS STIMULATION ON GROWTH.

There now remains the very interesting question as to whether the effect of long ether waves induce any variation of growth. The results given below show that growing plants not only perceive but respond to the stimulus of electric waves. The effects to be presently described are exhibited by all plants.

I shall, however, content myself in describing a typical experiment carried with the seedling of wheat. The specimen was mounted on the Balanced Crescograph, and the growth exactly balanced. This gives a horizontal record; an acceleration of growth above the normal is, in the

* "Comparative Electro-Physiology"—p. 149.

following records, represented by a down curve, and a retardation by an up-curve.

Effect of feeble stimulus: Experiment 161.—I first studied the effect of feeble stimulus. This was secured by decreasing the energy of sparks of the radiator. The response was an acceleration of rate of growth as seen in figure 156a. The analogy of this with the accelerating effect of sub-minimal intensity of light (p. 224) is very remarkable.

Effect of strong stimulus: Experiment 162.—The maximum energy radiated by my transmitter, as stated before, was only moderate. In spite of this its effect on plants was exhibited in a very striking manner. The balance was

FIG. 156.—Record of responses to electric wave by the Balanced Crescograph (*a*) response to feeble stimulus by acceleration of growth, (*b*) response to strong stimulus by retardation, (*c*) responses to medium stimulation—retardation followed by recovery. Down-curve represents acceleration, and up-curve retardation of growth: (Seedling of wheat.)

immediately upset, indicating a retardation of the rate of growth. The latent period, *i.e.*, the interval between the incident wave and the response, was only a few seconds (Fig. 156b). The record given in the figure was obtained with the moderate magnification of 2,000 times only. But with my Magnetic Crescograph, the magnification can easily be raised ten million times; and the response of plant to the space signalling can be exalted in the same proportion.

Under an intensity of stimulus slightly above the sub-minimal, the responses exhibit retardation of growth followed by quick recovery, as seen in the series of records given in Fig. 156c.

A remarkable peculiarity in the response was noticed during the course of the experiments. Strong stimulation by ether waves gives rise, as we have seen, to a very marked retardation of the rate of growth. Repeated stimulation induces fatigue, and temporary insensitiveness of the organ. Under moderate fatigue the effect is a prolongation of the latent period. Thus in a particular experiment the plant failed to give any response to a short signal. But after an interval of five minutes a marked response occurred to the wireless stimulus that had been received previously. The plant had perceived the stimulus but on account of fatigue the latent period was prolonged, from the normal 5 seconds to as many minutes.

SUMMARY.

Plants not only perceive, but also respond to long ether waves employed in signalling through space.

Mechanical response to wireless stimulation is exhibited by the leaf of *Mimosa pudica*.

All plants give electric response to the stimulus of long ether waves.

Growing plants exhibit response to electric waves by modification of rate of growth. Feeble stimulus induces an acceleration, while strong stimulus causes a retardation of the rate of growth.

The perceptive range of the plant is far greater than ours; it not only perceives but responds to the different rays of the vast ethereal spectrum.

XXXIX.—GEOTROPISM

By

SIR J. C. BOSE.

NO phenomenon of tropic movement appears so inexplicable as that of geotropism. There are two diametrically opposite effects induced by the same stimulus of gravity, in the root a movement downwards, and in the shoot a movement upwards. The seeming impossibility of explaining effects so divergent by the fundamental reaction of stimulus, has led to the assumption that the irritability of stem and root are of opposite character. I shall, however, be able to show that this assumption is unnecessary.

The difficulty of relating geotropic curvature to a definite reaction to stimulus is accentuated by the fact that the direction of the incident stimulus, and the side which responds effectively to it are not clearly understood ; nor is it known, whether the reaction to this stimulus is a contraction, or its very opposite, an expansion.

Taking the simple case of a horizontally laid shoot, the geotropic up-curvature is evidently due to differential effect of the stimulus on upper and lower sides of the organ. The up-curvature may be explained by one or the other of two suppositions : (1) that the stimulus of gravity induces contraction of the upper side ; or (2) that the fundamental reaction is not a contraction but an expansion and this of the lower side. The second of these two assumptions has found a more general acceptance.

Tropic curvatures in general are brought about by the differential effect of stimulus on two sides of the organ

Thus light falling on one side of a shoot induces local contraction, the rays being cut off from acting on the further side by the opacity of the intervening tissue. But there is no opaque screen to cut off the vertical lines of gravity,* which enter the upper side of a horizontally laid shoot and leave it by the lower side. Though lines of force of gravity are transmitted without hindrance, yet a differential action is found to take place, for the upper side, where the lines of force enter, becomes concave, while the lower side where they emerge becomes convex. Why should there be this difference ?

For the removal of various obscurities connected with geotropism it is therefore necessary to elucidate the following :

1. The sign of excitation is, as we found, a contraction and concomitant galvanometric negativity. Does gravitational stimulus, like stimulus in general, induce this excitatory reaction ?

2. What is the effective direction of geotropic stimulus ? In the case of light, we are able to trace the rays of light which is incident on the proximal side and measure the angle of inclination. In the case of gravity, the invisible lines of force enter by one side of the organ and leave by the other side. Assuming that the direction of stimulus is coincident with the vertical lines of gravity, is it the upper or the lower side of the organ that undergoes effective stimulation ?

3. What is the law relating to the 'directive angle' and the resulting geotropic curvature ? By the directive angle (sometimes referred to as the angle of inclination)

* I shall in what follows take the *direction* of vertical lines of gravity as that of movement of falling bodies, from above towards the centre of the earth.

is meant, as previously explained, the angle which direction of stimulus makes with the responding surface.

4. We have finally to investigate, whether the assumption of opposite irritabilities of the root and the shoot is at all justifiable. If not, we have to find the true explanation of the opposite curvatures exhibited by the two types of organs.

Of these the first three are inter-related. They will, however, be investigated separately; and each by more than one method of inquiry. The results will be found to be in complete harmony with each other.

I propose in this and in the following chapters to carry out the investigations sketched above, employing two independent methods of enquiry, namely, of mechanical and of electrical response. I shall first describe the automatic method I have been able to devise, for an accurate and magnified record of geotropic movement and its time relations.

THE GEOTROPIC RECORDER.

The recorder shown in figure 157 is very convenient for study of geotropic movement. The apparatus is four-sided and it is thus possible to obtain four simultaneous records with different specimens under identical conditions. The recording levers are free from contact with the recording surface. By an appropriate clock-work mechanism, the levers are pressed for a fraction of a second against the recording surfaces. The successive dots in the record may, according to different requirements, be at intervals varying from 5 to 20 seconds. The records therefore not only give the characteristic curves of geotropic movements of different plants, but also their time ations. For high magnification, I employ an Oscillating

Recorder, the short arm of the lever being 2·5 mm., and the long arm 250 mm., the magnification being a

FIG. 157.—The Quadruplex Geotropic Recorder.

hundredfold ; half that magnification is, however, sufficient for general purposes.

DETERMINATION OF THE CHARACTER OF GEOTROPIC REACTION.

The observed geotropic concavity of the upper side of a horizontally laid shoot may be due to excitatory *contraction* of that side, or it may result from passive yielding to the active responsive *expansion* of the lower side. The crucial test of excitatory reaction under geotropic stimulus is furnished by investigations on geo-electric response. When a shoot is displaced from vertical to horizontal position, *the upper side of the organ is found to undergo an excitatory electric change of galvanometric negativity* indicative of diminution of turgor and *contraction.* The electric change induced on the lower side is one of galvanometric positivity, which indicates an increase of turgor and expansion. The tropic effect of geotropic stimulus is thus similar to that of any other mode of stimulation, *i.e.*, a contraction of the upper (which in the present case is the proximal) and expansion of the lower or the distal side. The vertical lines of gravity impinge on the upper side of the organ which undergoes effective stimulation.

In order to show that the concavity of the upper side is not due to the passive yielding to the expansion of the lower half, I restrained the organ from any movement. I have explained that excitatory electric response is manifested even in the absence of mechanical expression of excitation ; and under geotropic stimulus, the securely held shoot gave the response of galvanometric negativity of the upper side. Hence the fundamental reaction under geotropic stimulus is excitatory *contraction* as under other modes of stimulation.

Finally, I employed the additional test of induced paralysis by application of intense cold. Excitatory physiological reaction is, as we know, abolished temporarily by the action of excessive cold.

Experiment 163.—I obtained records of mechanical response to determine the side which undergoes excitation under geotropic stimulus, the method of discrimination being local paralysis induced by cold. I took the flower-scapes of *Amarayllis* and of *Uriclis*, and holding them vertical applied fragments of ice on one of the two sides. I then laid the scape horizontal, first with cooled side below, the record showed that this did not affect the geotropic movement. But on cooling the upper side, the geotropic move-

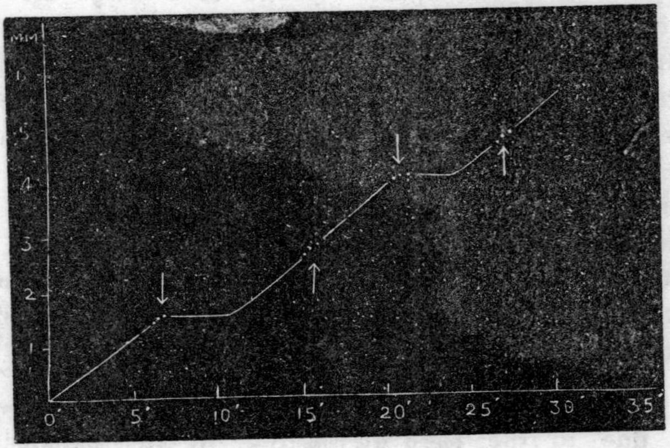

FIG. 158.—Effect of alternate application of cold on the upper and lower sides of the organ. Application of cold on upper side (down-pointing arrow) induces arrest of geotropic movement. Application below (up-pointing arrow) causes no arrest.

ment became arrested, and it was not till the plant had assumed the temperature of the surroundings that the geotropic movement became renewed. Figure 158 shows the effect of alternate application of cold, on the upper and lower sides of the organ.* In the shoot, therefore, it is the upper side of the organ that becomes effectively stimulated. Before proceeding further I shall make brief reference to the highly suggestive statolithic theory of gravi-perception.

THEORY OF STATOLITHS.

With regard to the perception of geotropic stimulus there can be no doubt that this must be due to the

* "Plant Response"— p. 505.

effect of weight of cell contents, whether of the sap itself, or of the heavy particles contained in the cells, exerting pressure on the sensitive plasma. The theory of statoliths advocated by Noll, Haberlandt and Nemec (in spite of certain difficulties which further work may remove) is the only rational explanation hitherto offered for gravi-perception. The sensitive plasma is the ectoplasm of the entire cell, and statoliths are relatively heavy bodies, such as crystals and starch grains. Haberlandt has found statoliths in the apo-geotropic organs like stems.* When the cell is laid horizontal, it is the lower tangential wall which has to support the greater weight, and thus undergo excitation. In the case of multicellular plants laid horizontally, the excitation on the upper side is, as we have seen, the more effective than on the lower side. This inequality, it has been suggested, is probably due to this difference that the statoliths on the upper side press on the inner tangential walls of the cells while those on the lower side rest on the outer tangential walls.

When the organ is held erect, the action of statoliths would be symmetrical on the two sides. But when it is laid horizontal a complete rearrangement of the statoliths will take place, and the differential effects on the upper and lower sides will thus induce geotropic reaction. This *period of migration* must necessarily be very short; but the reaction time, or the latent period, is found to be of considerable duration. "Even in rapidly reacting organs there is always an interval of about one to one and a half hours, before the horizontally placed organ shows a noticeable curvature, and this latent period may in other cases be extended to several hours."† This great difference between the *period of migration* and the *latent*

* Haberlandt—"Physiological Plant Anatomy"—p. 603.

† Jost—*Ibid*, p. 437.

period offers a serious difficulty in the acceptance of the theory of statoliths. But it may be urged that the latent period has hitherto been obtained by relatively crude methods, and I therefore undertook a fresh determination of its value by a sensitive and accurate means of record.

DETERMINATION OF THE LATENT PERIOD.

As regards the interpretation of the record of geotropic movement, it should be borne in mind that after the perception of stimulus a certain time must elapse before the induced growth-variation will result in curvature. There is again another factor which causes delay in the exhibition of true geotropic movement; for the up-movement of stems, in response to the stimulus of gravity, has to overcome the opposite down movement, caused by weight, before it becomes at all perceptible. On account of the bending due to weight there is a greater tension on the upper side, which as we have seen (p. 193), enhances the rate of growth, and thus tends to make that side convex. The exhibition of geotropic response by induced contraction of the excited upper side thus becomes greatly delayed. In these circumstances I tried to discover specimens in which the geotropic action would be quick, and in which the retarding effect of weight could be considerably reduced.

Geotropic response of flower stalk of Tuberose : Experiment 164.—For this I took a short length of flower stalk of tuberose in a state of active growth; the flower head itself was cut off in order to remove unnecessary weight. After a suitable period of rest for recovery from the shock of operation, the specimen was placed in a horizontal position, and its record taken. The successive dots in the curve are at intervals of 20 seconds, and

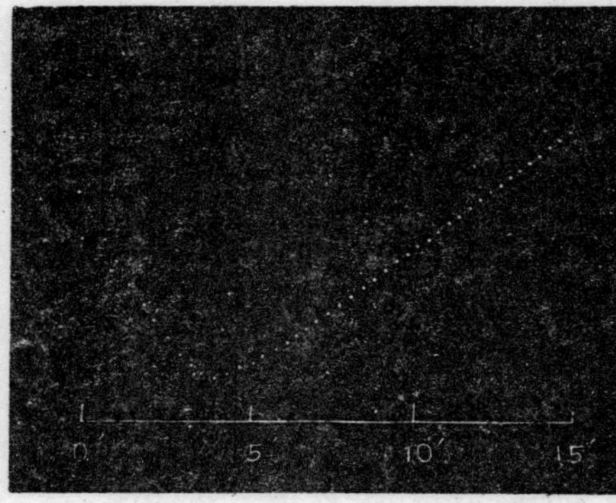

FIG. 159.

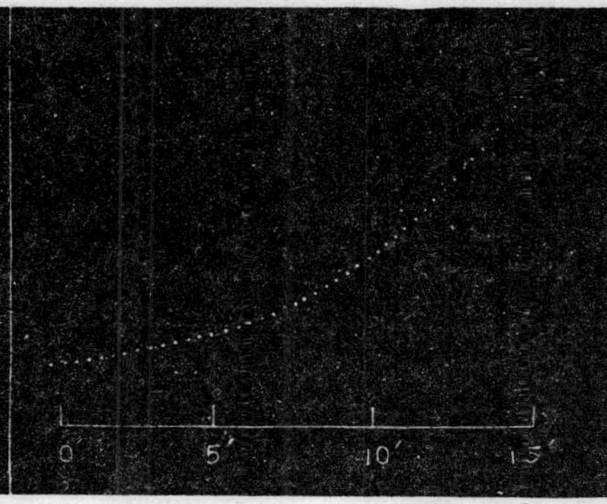

FIG. 160.

FIG. 159.—Geotropic response of flower stalk of tube rose: preliminary down-movement is due to weight.

FIG. 160.—Geotropic response of petiole of *Tropæolum*: latent period shorter than 20 seconds.

the geotropic up-movement is seen to be initiated (Fig. 159) after the tenth dot, the latent period being thus 3 minutes and 20 seconds, the greater part of which was spent in overcoming the down-movement caused by the weight of the organ.

Geotropic response of petiole of Tropæolum: *Experiment 165.*—I expected to obtain still shorter latent period by choosing thinner specimens with less weight. I therefore took a cut specimen of the petiole of *Tropæolum*, and held it at one end. The lamina was also cut off in order to reduce the considerable leverage exerted by it. The response did not now exhibit any preliminary down-movement, and the geotropic up-movement was commenced within a few seconds after placing the petiole in a horizontal position (Fig. 160). The successive dots in the record are at intervals of 20 seconds and the second dot already exhibited an up-movement; the latent period is therefore shorter than 20 seconds. It will thus be seen that the latent period in this case is of the same order as the hypothetical period of migration of the statoliths.

I may state here that I have been successful in devising an electric method for the determination of the latent period, in which the disturbing effect of the weight of the organ is completely eliminated. Applying this perfect method, I found that the latent period was in some cases as short as a second. The experiment will be found fully described in a later chapter.

THE COMPLETE GEOTROPIC CURVE.

The characteristics of the geotropic curve are similar to those of other tropic curves. That is to say the susceptibility for excitation is at first feeble; it then increases at a rapid rate; in the third stage the rate becomes uniform; and finally the curvature attains a maximum value and the organ attains a state of geotropic equilibrium (cf. page 353). The period of completion of the curve varies in different specimens from a few to many hours.

Experiment 166.—The following record was obtained with a bud of *Crinum*, the successive dots being at intervals of 10 minutes. After overcoming the effect of weight (which took an hour), the curve rose at first slowly, then rapidly. The period of uniformity of movement is seen to be attained after three hours and continued for nearly 90 minutes. The final equilibrium was reached after a period of 8 hours (Fig. 161).

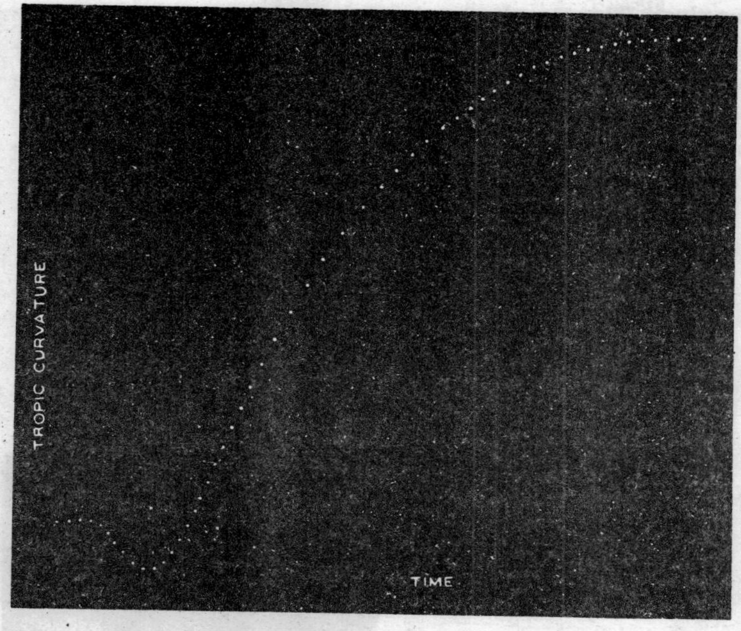

FIG 161.—The Complete Geotropic curve (*Crinum*).

For studying the effect of an external agent on geotropic action, the period of uniform movement is the most suitable. Acceleration of the normal rate (with enhanced steepness of curve) indicates that the external agent acts with geotropism in a concordant manner; depression of the rate with resulting flattening of the curve shows, on the other hand, the antagonistic effect of the outside agent.

DETERMINATION OF EFFECTIVE DIRECTION OF STIMULUS.

The experiments which have been described show that it is the upper side (on which the vertical lines of gravity impinge) that undergoes excitation. The vertical lines of gravity must therefore be the direction of incident stimulus. This conclusion is supported by results of three independent lines of inquiry: (1) the algebraical summation of effect with that of a different stimulus whose direction is known, (2) the relation between the directive angle and geotropic reaction, and (3) the torsional response under geotropic stimulus.

EFFECT OF ALGEBRAICAL SUMMATION.

Experiment 167.—A flower bud of *Crinum* is laid horizontally, and record taken of its geotropic movement. On application of light on the upper side at L,

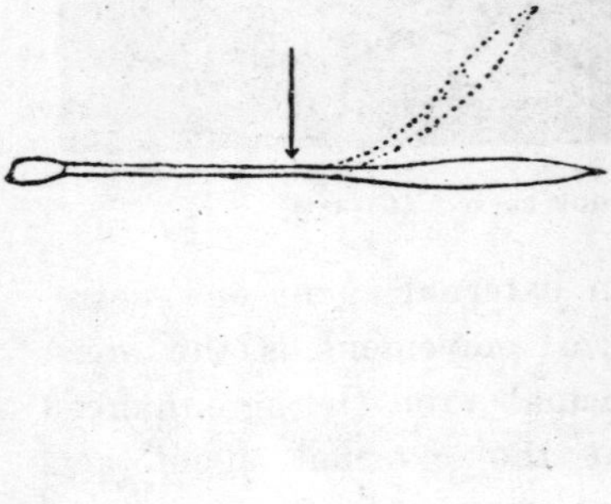

FIG. 162.

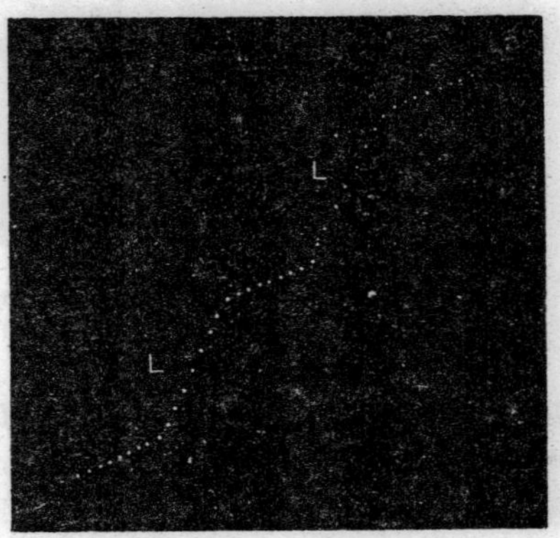

FIG. 163.

FIG. 162.—Stimulus of light or gravity, represented by arrow, induces up curvature as seen in dotted figure.

FIG. 163.—The effect of super-imposition of photic stimulus. The first, third, and fifth parts of the curve, give normal record under geotropic stimulus. Rate, of up-movement enhanced under light L.

the responsive movement is enhanced, proving that gravity and light are inducing similar effects. On the cessation of light, the original rate of geotropic movement is restored (Fig. 163). Application of light of increasing intensity from below induces, on the other hand, a diminution, neutralisation, or reversal of geotropic movement.

Light acting vertically from above induces a concavity of the excited upper side in consequence of which the organ moves, as it were, to meet the stimulus. The geotropic response is precisely similar. In figure 162 the arrow represents the direction of stimulus which may be rays of light or vertical lines of gravity.

ANALOGY BETWEEN THE EFFECTS OF STIMULUS OF LIGHT AND OF GRAVITY.

In geotropic curvature we may for all practical purposes regard the direction of stimulus as coinciding with the vertical lines of gravity. The analogy between the effects of light and of gravity is very close* ; in both the induced curvature is such that the organ moves so as to meet the stimulus. This will be made still more evident in the investigations on torsional geotropic response described in a subsequent chapter. The tropic curve under geotropic stimulus is similar to that under photic stimulus. The tropic reaction, both under the stimulus of light and of gravity, increases similarly with the 'directive' angle. These real analogies are unfortunately obscured by the use of arbitrary terminology used in description of the geotropic curvature of the shoot. In figure 163 records are given of the effects of vertical light and of vertical stimulus of gravity, on the responses of the horizontally laid bud of *Crinum*. In both, the upper side undergoes contraction and the movement of response carries the

* Exception to this will be found in page 336, where explanation is offered for the difference.

organ upwards so as to place it parallel to the incident stimulus. Though the reactions are similar in the two cases, yet the effect of light is termed *positive* phototropism, that of gravity *negative* geotropism. I would draw the attention of plant-physiologists to the anomalous character of the existing nomenclature. Geotropism of the shoot should, for reasons given above, be termed *positive* instead of *negative*, and it is unfortunate that long usage has given currency to terms which are misleading, and which certainly has the effect of obscuring analogous phenomena. Until the existing terminology is revised, it would perhaps be advisable to distinguish the geotropism of the shoot as *Zenithotropism* and of the root as *Nadirotropism*.

RELATION BETWEEN THE DIRECTIVE ANGLE AND GEOTROPIC REACTION.

When the main axis of the shoot is held vertical, the angle made by the surface of the organ with lines of force of gravity is zero, and there is no geotropic effect. The geotropic reaction increases with the directive angle; theoretically the geotropic effect should vary as the sine of the angle. I shall in the next chapter describe the very accurate electrical method, which I have been able to devise for determination of relative intensities of geotropic action at various angles. Under perfect conditions of symmetry, the intensity of effect is found to vary as the sine of the directive angle. This quantitative relation fully demonstrates that geotropic stimulus acts in a definite direction which coincides with the vertical lines of gravity.

The conditions of perfect symmetry for study of geotropic action at various angles will be fully described in the next chapter. In the ordinary method of experimentation with mechanical response the organ is rotated in a vertical plane. The geotropic movement is found increased as the directive angle is increased from zero to 90°.

DIFFERENTIAL GEOTROPIC EXCITABILITY.

It has been shown that geotropic stimulus acts more effectively on the upper side of the organ. The intensity of geotropic reaction is, moreover, modified by the excitability of the responding tissue. It is easy to demonstrate this by application of depressing agents on the more effective side of the organ. The rate of geotropic up-movement will be found reduced, or even abolished by the local application of cold, anæsthetics like chloroform, and of poisonous potassium cyanide solution.

The different sides of a dorsiventral organ are unequally excitable to different forms of stimuli. I have already shown (p. 85) that the lower side of the pulvinus of *Mimosa*, is about 80 times more excitable to electric stimulus than the upper side. Since the effect of geotropic stimulus is similar to that of other forms of stimuli, the lower side of the pulvinus should prove to be geotropically more excitable than the upper side. This I have been able to demonstrate by different methods of investigation which will be described in the following chapters.

Under ordinary circumstances, the upper half of the pulvinus is, on account of its favourable position, more effectively stimulated by geotropic stimulus; in consequence of this the leaf assume a more or less horizontal position of "dia-geotropic" equilibrium. But when the plant is inverted the more excitable lower half of the organ now occupies the favourable position for geotropic excitation.

The leaf now erects itself till it becomes almost parallel to the stem. The response of the same pulvinus which was formerly "dia-geotropic" now becomes "negatively geotropic"; but an identical organ cannot be supposed to possess two different specific sensibilities. The normal horizontal position assumed by the leaf is, therefore, due to differential geotropic excitabilities of the two sides of a dorsiventral organ.

I have explained (p. 401) that when the pulvinus of *Mimosa* is subjected to lateral stimulation of any kind, it undergoes a torsion, in virtue of which the less excitable half of the organ is made to face the stimulus. Experiments will be described in a subsequent chapter which show that geotropic stimulus also induces similar torsional response. The results obtained from this method of enquiry give independent proof: (1) that the lower half of the pulvinus is geotropically the more excitable, and (2) that the direction of incident geotropic stimulus is the vertical line of gravity which impinges on the upper surface of of the organ.

SUMMARY.

The stimulus of gravity is shown to induce an excitatory reaction which is similar to that induced by other forms of stimulation. The direct effect of geotropic stimulus is an incipient contraction and retardation of rate of growth.

The upper side of a horizontally laid shoot is more effectively stimulated than the lower side, the excited upper side becoming concave. Electrical investigation also shows that it is the upper side that undergoes direct stimulation.

Tropic reactions are said to be positive, when the directly stimulated side undergoes contraction with the result that the organ moves to meet the stimulus. According to this test, the geotropic response of the stem is *positive.*

The geotropic response is delayed by the bending down of the horizontally laid shoot. Reduction of weight is found to shorten the latent period; in the case of the petiole of *Tropæolum* this is shorter than 20 seconds. The latent period of geotropic response is found to be of the same order as the "migration period" of the hypothetical statoliths.

The complete geotropic curve shows characteristics which are similar to tropic curves in general.

In a dorsiventral organ the geotropic excitabilities of the upper and lower sides are different. In the pulvinus of *Mimosa* the geotropic excitability of the lower half is greater than that of the upper half. The differential excitabilities of a dorsiventral organ modifies its position of geotropic equilibrium.

XL.—GEO-ELECTRIC RESPONSE OF SHOOT

By

SIR J. C. BOSE,

Assisted by

SATYENDRA CHANDRA GUHA, M.SC.

THE experiments that have been described in the preceding chapter show that the upper side of a horizontally laid shoot undergoes excitatory contraction, in consequence of which the organ bends upwards. The fundamental geotropic reaction is, therefore, not expansion, but contraction which results from all modes of stimulation.

In confirmation of the above, I wished to discover and employ new means of detecting excitatory reaction under geotropic stimulus. In regard to this, I would refer to the fact which I have fully established that the state of excitation can be detected by the induced electromotive change of galvanometric negativity. This electrical indication of excitation may be observed even in plants physically restrained from exhibiting response by mechanical movement.*

ELECTRIC RESPONSE TO STIMULUS.

Before giving account of the results of investigations on the detection of geotropic excitation by means of electric response, I shall describe a few typical experiments which will fully explain the method of the electrical investigation, and show the correspondence of mechanical and electric responses. I have explained how tropic curvatures are brought about by the joint effects, of

* "Comparative Electro-Physiology." p. 20.

contraction of the directly excited proximal side A, and the expansion of the distal side B. In the diagram of mechanical response to stimulus (Fig. 164a) the excitatory contraction is indicated by – sign, and the expansion, by + sign. The resulting movement is, therefore, towards the stimulus as shown by the curved arrow.

I shall now describe the corresponding electric effects in response to unilateral stimulus. We have to determine the induced electrical variation at the proximal side A, and at the distal side B.

Electric response to direct stimulation : Experiment 168.— For the determination of electric response at the directly excited proximal side A, we take a shoot with a lateral leaf. The point A, which is to undergo stimulation, is connected with

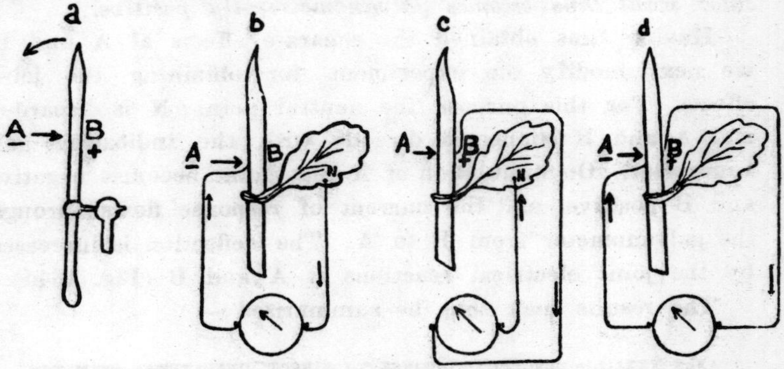

FIG. 164.—Diagrammatic representation of the mechanical and electrical response to direct unilateral stimulation indicated by arrow :—

(*a*) Positive mechanical response (curved arrow) due to contraction of directly stimulated A, and expansion of indirectly stimulated B.

(*b*) Electric response of induced galvanometric negativity of A under direct stimulation

(*c*) Electric response of induced galvanometric positivity at the distal point B.

(*d*) Additive effects of direct and indirect stimulations ; galvanometric negativity of the directly stimulated proximal A, and galvanometric positivity of the indirectly stimulated distal point B.

one terminal of the galvanometer, the other terminal being led to an indifferent or neutral point N on the leaf.

Application of any form stimulus at A, gives rise to an electric current which flows through the galvanometer from the neutral to the excited point A (Fig. 164b). *The directly stimulated point A thus becomes galvanometrically negative.* The "action" current lasts during the application of stimulus and disappears on its cessation.

Electric response to indirect stimulation: Experiment 169.—We have also seen that application of stimulus at A causes indirect stimulation of the distal point B resulting in an increase of turgor and expansion. The corresponding electric change of the indirectly stimulated point B is found in the responsive current, which flows now through the galvanometer from the indirectly stimulated B to the neutral point N (Fig. 164c). *The indirectly stimulated point thus becomes galvanometrically positive.*

Having thus obtained the separate effects at A and B, we next modify the experiment for obtaining the joint effects. For this purpose the neutral point N is discarded and A and B connected directly with the indicating galvanometer. On stimulation of A that point becomes negative and B positive, and the current of response flows through the galvanometer from B to A. The deflection is increased by the joint electrical reactions at A and B (Fig. 164d).

The results may thus be summarised :—

TABLE XXXIII.—ELECTRIC RESPONSE TO DIRECT UNILATERAL STIMULUS.

Electrical change at the proximal side A.	Electrical change at the distal side B.
Galvanometric negativity indicative of contraction and diminution of turgor.	Galvanometric positivity indicative of expansion and increase of turgor.
The corresponding tropic curvature is positive movement towards stimulus.	

Galvanometric negativity is thus seen to indicate the effect of direct stimulus, and galvanometric positivity that of indirect stimulus. We thus see the possibility of electric detection of the effects of geotropic stimulation. This method would, moreover, enable us to discriminate the side of the organ which undergoes greater excitation.

EXPERIMENTAL ARRANGEMENTS FOR OBTAINING GEO-ELECTRIC RESPONSE.

Returning to the investigation on electric response to geotropic stimulus, the specimen of plant is at first held erect; two electrodes connected with a sensitive galvanometer are applied, one to an indifferent point, and the other to one side of the shoot. The sensitiveness of the galvanometer was such that a current of one millionth of an ampere produced a deflection of the reflected spot of light through 1,000 divisions of the scale. An action current is produced on displacement of the plant from vertical to horizontal position.

Non-polarisable electrodes.—The electrical connections with the plant are usually made by means of non-polarisable electrodes (amalgamated zinc rod in zinc-sulphate solution and kaolin paste with normal saline). I at first used this method and obtained all the results which will be presently described. But the employment of the usual non-polarisable electrodes with liquid electrolyte is, for our present purpose, extremely inconvenient in practice; for the plant-holder with the electrodes has to be rotated from vertical to horizontal through 90°. The reliability of the non-polarisable electrode, moreover, is not above criticism. The zinc-sulphate solution percolates through the kaolin paste and ultimately comes in contact with the plant, and seriously affects its excitability. The name non-polarisable electrode is in reality a misnomer;

for the action current (whose polarising effect is to be guarded against) is excessively feeble, being of the order of a millionth of an ampere or even less; the counter polarisation induced by such a feeble current is practically negligible.

The idea that non-polarisable electrodes are meant to ged rid of polarisation is not thus justified by the facts of the case. The real reason for its use is very different; the electrical connections with the plant has to be made ultimately by means of two metal contacts. If we take two pieces of metal even from the same sheet, and put them in connection with the plant, a voltaic couple is produced owing to slight physical differences between the two electrodes. Amalgamation of the two zinc rods with mercury reduces the electric difference but cannot altogether eliminate it.

I have been able to wipe off the difference of potential between two pieces of the same metal, say of platinum, and by immersing them in dilute salt solution from a voltaic couple. The circuit is kept complete for 24 hours, and the potential of the two electrodes by this process is nearly equalised. A perfect equality is secured by repeated warming and cooling of the solution and by sending through the circuit, alternating current which is gradually reduced to zero. I have by this means been able to obtain two electrodes which are iso-electric. The specially prepared electrodes (made of gold or platinum wire) are put in connection with the plant through kaolin paste moistened with normal saline solution. Care should be taken to use opaque cover over the plant-holder, so as to guard against any possible photo-electric action; moistened blotting paper maintains the closed chamber in a uniform humid condition.

The direct method of contact described above is extremely convenient in practice; the resistance of contact is considerably reduced, and there is no possibility of its variation during the necessary process of rotation of the plant for subjecting it to geotropic action.

GEO-ELECTRIC RESPONSE OF THE UPPER AND LOWER SIDES OF THE ORGAN.

We have next to discover the electric change induced by geotropic stimulus on the upper and lower sides of the organ. For this purpose it is necessary to find a neutral point which is not affected by the inclination of the organ from vertical to horizontal position. For the present experiment, I employed the flower of the water lily *Nymphæa*, the peduncle of which is sensitive to geotropic action. One electrical contact is made with a

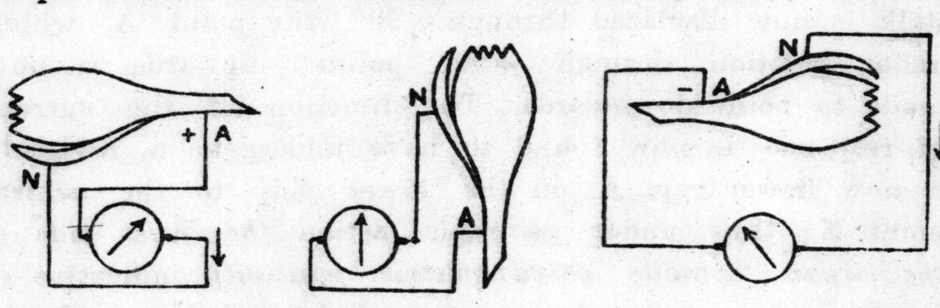

FIG. 165.—Diagrammatic representation of geo-electric response. The middle figure represents vertical position. In figure to the right rotation through +90° has placed A above with induced electric change of galavanometric negativity of A. In the figure to the left, rotation is through -90° A being below; the electric response is by induced galvanometric positivity of A. For simplification of diagram, vertical position of sepal is not always shown in the figure.

sepal, which is always kept vertical; the other electric contact is made at the point A, on one side of the flower stalk (Fig. 165). On making connections with a sensitive galvanometer a very feeble current was found, which was due to slight physiological difference between the neutral point, N, and A. This natural current may be allowed to remain, the action current due to geotropism being *superposed on it;* or the natural current may be neutralised by means of a potentiometer and the reflected spot of light brought to zero of the scale.

Induced electric variation on upper side of the organ: Experiment 170.—While the sepal is held vertical, the stalk is displaced through +90° so that the point A is above. Geotropic stimulation is at once followed by a responsive current which flows through the galvanometer from N to A, the upper side of the organ thus exhibiting excitatory reaction of galvanometric negativity (Right-hand figure of 166). When the stalk is brought back to vertical position geotropic stimulation disappears, and with it the responsive current.

Electric response of the lower side: Experiment 171.—The stalk is now displaced through −90°; the point A, which under rotation through +90° pointed upwards, is now made to point downwards. The direction of the current of response is now found to have undergone a reversal; it now flows from A on the lower side to the neutral point N; thus under geotropic action *the lower side of the organ exhibits galvanometric positivity* indicative of increase of turgor and expansion (Left-hand figure 166).*

Having thus found that the upper side of the organ under geotropic stimulus becomes galvanometrically negative, and the lower side, galvanometrically positive, we make electric connections with two diametrically opposite points of the shoot A and B, and subject the organ to alternate rotation through +90° and −90°. The electromotive changes induced at the two sides now became algebraically summated. I employ two methods for geotropic

* For detailed account cf. Chapter XLIII.

stimulation: that (1) of Axial Rotation, and (2) of Vertical Rotation.

METHOD OF AXIAL ROTATION.

In the method of Axial Rotation, the organ is held with its long axis horizontal (Fig. 166 H). We have seen that the geotropic action increases with the angle which the responding surface of the organ makes with the

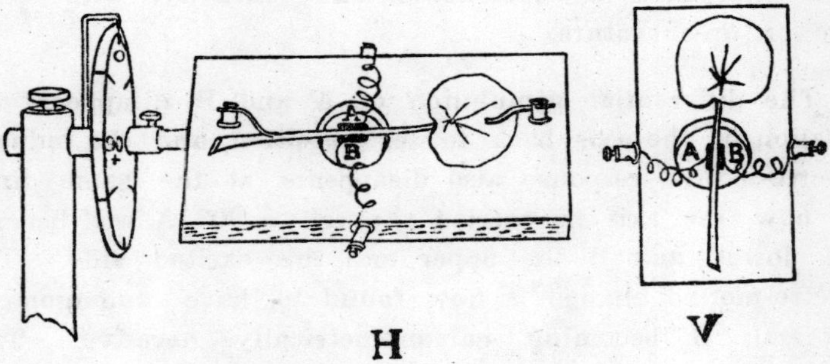

FIG. 166.—Diagrammatic representation of the Method of Axial Rotation H, and of Vertical rotation V (see text).

vertical lines of gravity. When the organ is held with its length horizontal, the angle made by its two sides, A and B, with the vertical is zero and there is thus no geotropic effect. There is, moreover, no differential effect, since the two sides are symmetrically placed as regards the vertical lines of force. The plant is next rotated round its long axis, the angle of rotation being indicated in the circular scale. When the rotation is through +90 , A is above and B below; this induces a differential geotropic effect, the upper side exhibiting excitatory electric change of galvanometric negativity.

Experiment 172.—I shall, as a typical example, give a detailed account of experiments with the petiole of *Tropæolum* which was found so highly excitable to geotropic stimulus (p. 434). The specimen was held horizontal with two symmetrical contacts at the two sides, the electrodes being connected in the usual manner with the indicating galvanometer. When the plant is rotated through +90° there is an immediate current of response, the upper side becoming ***galvanometrically negative.*** This excitatory reaction on the upper side finds, as we have seen, mechanical expression by contraction and concavity, with positive or up-curvature.

The differential stimulation of A and B disappears on rotation of the axis back to zero position, and the induced electro-motive response also disappears at the same time. If now the axis be rotated through − 90°, A will become the lower, and B the upper and the excited side. The electro-motive change is now found to have undergone a reversal, B becoming galvanometrically negative. This induced electro-motive variation under geotropic stimulus is

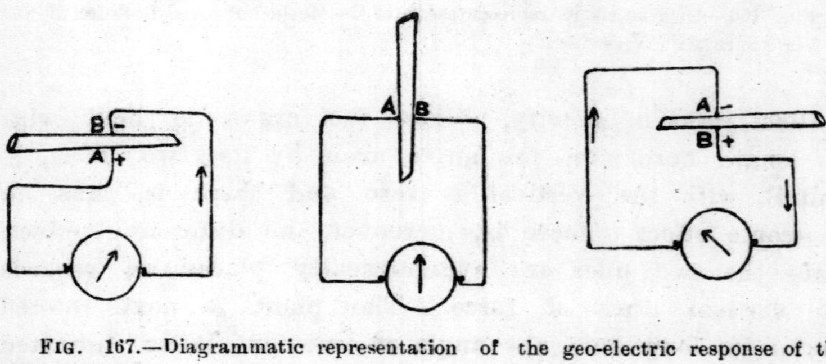

FIG. 167.—Diagrammatic representation of the geo-electric response of the shoot (see text).

of considerable intensity often exceeding 15 millivolts. The characteristic electric change is shown diagrammatically in figure 167 in which the middle figure shows the symmetrical or zero position. On rotation through + 90° (figure to the right) A occupies the upper and B the lower position. A is seen to exhibit induced change of galvanometric negativity. Rotation through – 90° reverses the current of response, as B now occupies the upper and A the lower position.

CHARACTERISTICS OF GEO-ELECTRIC RESPONSE.

There are certain phenomena connected with the electric response under geotropic stimulus which appear to be highly significant. According to statolithic theory

"Geotropic response begins as soon as an organ is deflected from its stable position, so that a few starch-grains press upon the ectoplasts occupying the walls which are underneath in the new position; an actual rearrangement of the starch-grains is therefore not an essential condition of stimulation. As a matter of fact, the starch-grains do very soon migrate on to the physically lower walls, when a positively or negatively geotropic organ is placed horizontally, with the result that the intensity of stimulation gradually increases attaining its maximum value when all the falling starch-grains have moved on to the lower region of the ectoplast. The time required for the complete rearrangement of the statoliths may be termed the period of migration; its average length varies from five to twenty minutes in different organs."*

Stimulation, according to the statolithic theory, is induced by the displacement of the particles. The diameter of the geotropically sensitive cells is considerably less than 0·1 mm.; and the stimulus will be perceived after the very short interval taken by the statoliths to fall through a space shorter than 0·1 mm. This may be somewhat delayed by the viscous nature of the plasma, but in any case the period

* Haberlandt—*Ibid*—p. 598.

for perceptible displacement of the statoliths should be very short, about a second or so, and the latent period of perception of stimulus should be of this order.

The mechanical indication of response to stimulus is delayed by a period which is somewhat indefinite; for the initiation of responsive growth variation will necessarily lag behind the perception of stimulus.

Experiment 173.—The mechanical response with its drawbacks is thus incapable of giving an accurate value of the latent period. The electrical method of investigation labours under no such disadvantage, since the excitation is here detected even in the absence of movement. The perception of stimulus will thus be followed by response without undue delay. I shall in this connection give a record of electric response of the quickly reacting petiole of *Tropæolum*, when the angle of inclination is increased from zero

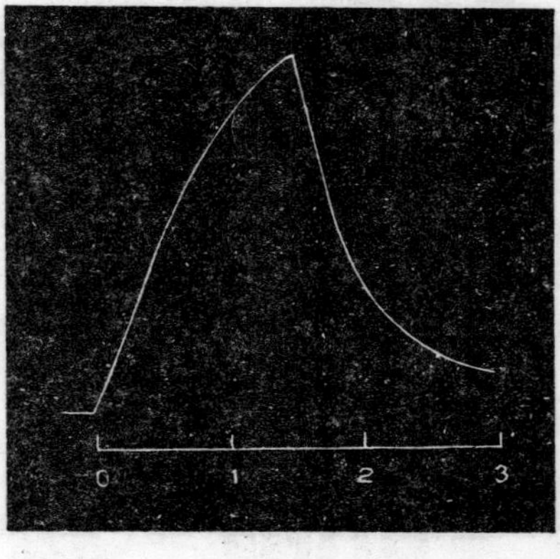

FIG. 168.—Geo-electric response of the petiole of *Tropæolum*.

to 90°. The responsive movement of the galvanometer spot of light was initiated in less than 5 seconds and the maximum deflection was reached in the course of 90 seconds. The angle was next reduced to zero, and the deflection practically disappeared in the further course of a minute and a half (Fig. 168). There was a small "excitation remainder". But with vigorous specimens the recovery is complete.

The latent period of quickly reacting petiole of *Tropæolum* is thus about 5 seconds, a value which is more consonant with the idea of particles inducing excitation by their fall through an exceedingly short distance. In very sluggish organs latent period may be as long as a minute (Fig. 169), which is considerably shorter than an

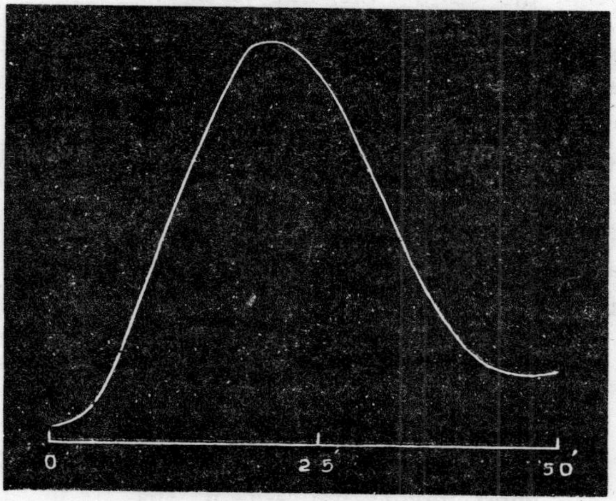

FIG. 169.—Geo-electric response of the scape of *Uriclis*.

hour, the generally accepted value. Further even in the electric response, the latent period will be delayed beyond the period of perception. For this perception takes place in some unknown sensitive layer in the interior of the tissue, while electric contact is made with the epidermis outside. It is obvious that certain time must elapse before the excitation, initiated at the sensitive layer, should reach the epidermis. Under ideal conditions of experiment which will be described in a subsequent chapter, the latent period for geotropic excitation, I find, to be sometimes as short as a second.

PHYSIOLOGICAL CHARACTER OF GEO-ELECTRIC RESPONSE.

The intensity of the electro-motive variation is found to depend on the physiological vigour of the specimen. The *Tropæolum* plant, used for most of the above experiments, are at the best condition of growth in Calcutta in February; after this the plants begin to decline in March and die off by the end of April.

Experiment 174.—In February the intensity of electric response was nearly double of that in March; it was only in March that I made quantitative determination of the induced electro-motive force between the upper and lower contacts on rotation of the specimen from zero to 90°. The E. M. F. was determined by the potentiometer method. I give below the following typical values obtained with two different specimens:—

Specimen	Induced E. M. F.
(1)	12 millivolts.
(2)	15 ,,

In the most favourable season the induced electro-motive force is likely to exceed the above value very considerably.

Effect of Age.—While a young petiole gave the above value, an old specimen from the same plant exhibited no response. The plants were in a dying condition in April and all indications of electrical reaction were found abolished. The physiological character of the response was also demonstrated by first obtaining the normal electric response in a vigorous specimen; after death, by immersion in boiling water, the plant gave no electric response to geotropic stimulus.

EFFECT OF DIFFERENTIAL EXCITABILITY OF THE ORGAN.

I have hitherto described the geo-electric effect of radial and isotropic organs. The induced E. M. F. at 90° was found practically the same whether A was above and B below, and *vice versâ.* In the mechanical response of the pulvinus of *Mimosa*, the geotropic excitability was, however, found to be greater in the lower half than in the upper (p. 440). I wished to investigate the question of differential geotropic excitability anew, by means of electric response.

Experiment 175.—Electric connections with the galvanometer were made with the upper and lower halves of the pulvinus, the organ being placed in the vertical or neutral position. The angle of inclination was then increased to 90° in the positive and negative directions alternately.

TABLE XXXIV.—DIFFERENCE OF GEO-ELECTRIC RESPONSE OF UPPER AND LOWER HALVES OF THE PULVINUS OF *Mimosa*.

Specimen.	Position of particular half of pulvinus.	Induced E. M. F.
(1)	Upper half above	23 millivolts.
	Lower half above	30 „
(2)	Upper half above	16 „
	Lower half above	29 „

In the former case the upper half of the pulvinus occupied the up-position; in the second case the up-position was occupied by the lower half of the pulvinus. In both cases strong electric responses were obtained, the upper point of contact being always galvanometrically negative. There was, however, a difference between the two responses, the excitatory electro-motive variation was invariably greater when the lower half of the organ occupied the favourable up-position. This will be seen from the results of two typical experiments in table given above.

The electrical mode of investigation thus leads to confirm the result obtained with mechanical method that the lower half of the pulvinus of *Mimosa* is geotropically more excitable than the upper half.

RELATION BETWEEN ANGLE OF INCLINATION AND GEOTROPIC EFFECT.

In the Method of Axial Rotation, the condition of the experiment is ideally perfect; in the neutral position the sides A and B are both parallel to the vertical lines of gravity, and are little affected by geotropic reaction. As the specimen is rotated on its long axis the vertical component of the force of gravity increases with the angle of inclination. The hypothetical statolithic particles will become displaced all along the cell, and the vertical pressure exerted by them will also increase with the angle.

The geo-electric response will then afford us a measure of the intensity of excitation induced at various angles of inclination. The mechanical response on account of its inherent defects does not afford us the true relation between the angle of inclination and intensity of geotropic reaction. But the electric method of inquiry is free from the defects of the mechanical method.

Experiment 176.—The specimen was rotated so that the angle of rotation was 45°, and the maximum electric response observed. The angle was next increased to 90° and the reading for the enhanced response taken. The ratio of the geo-electric response at 90° and 45°, thus affords us a measure of the effective stimulations at the two angles. I give below a table which gives results obtained with 24 different specimens.

TABLE XXXV—RELATION BETWEEN ANGLE OF INCLINATION AND GEOTROPIC EFFECT.

No. of specimen.	Galvanometric deflection.		Ratio $\frac{b}{a}$.
	(*a*) at 45°	(*b*) at 90°	
1 ...	70 divisions ...	110 divisions ...	1·5
2 ...	30 ,, ...	45 ,, ...	1·5
3 ...	90 ,, ...	126 ,, ...	1·4
4 ...	70 ,, ...	100 ,, ...	1·4
5 ...	21 ,, ...	33 ,, ...	1·6
6 ...	30 ,, ...	50 ,, ...	1·6
7 ...	12 ,, ...	20 ,, ...	1·6
8 ...	14 ,, ...	20 ,, ...	1·4
9 ...	10 ,, ...	16 ,, ...	1·6
10 ...	45 ,, ...	75 ,, ...	1·5
11 ...	25 ,, ...	40 ,, ...	1·6
12 ...	14 ,, ...	20 ,, ...	1·4
13 ...	13 ,, ...	20 ,, ...	1·5
14 ...	30 ,, ...	50 ,, ...	1·5
15 ...	38 ,, ...	54 ,, ...	1·4
16 ...	50 ,, ...	75 ,, ...	1·5
17 ...	55 ,, ...	90 ,, ...	1·5
18 ...	13 ,, ...	20 ,, ...	1·5
19 ...	17 ,, ...	25 ,, ...	1·4
20 ...	80 ,, ...	130 ,, ...	1·5
21 ...	15 ,, ...	22 ,, ...	1·4
22 ...	45 ,, ...	75 ,, ...	1·5
23 ...	135 ,, ...	220 ,, ...	1·6
24 ...	55 ,, ...	93 ,, ...	1·5
Mean ratio = 1·49			

The mean ratio 1·49 may thus be regarded as the relative geotropic effects at 90° and 45°; this is practically

...

the same as $\frac{\text{Sin } 90^\circ}{\text{Sin } 45^\circ}=1{\cdot}4$. Hence we arrive at the following law:

The intensity on geotropic action varies as the sine of the directive angle.

METHOD OF VERTICAL ROTATION.

I have hitherto described results obtained with the Method of Axial Rotation; I shall now take up the second method, that of Vertical Rotation, diagrammatic representation of which is given in figure 166V. The specimen is held vertical and two electrical contacts, A and B, made with the two lateral sides; it is then rotated round a horizontal axis perpendicular to the length of the specimen. Rotation may be carried in a right-handed direction with increasing angle with the vertical. The point A is thus subjected to enhanced geotropic stimulation and exhibits increasing electric change of galvanometric negativity; continuous decrease of angle of inclination to zero by rotation in the reverse direction causes a disappearance of the induced electric change. The rotation is next continued in the negative direction by which the point B is increasingly subjected to geotropic action. B is now found to exhibit excitatory reaction, the current of response having undergone a reversal. Rotation to the right and left will be distinguished by plus and minus signs.

ELECTRIC RESPONSE THROUGH AN ENTIRE CYCLE.

Experiment 177.—When the specimen is vigorous, characteristic response with its changing sign may be obtained through an entire cycle from 0° to + 45° to + 90°; then back to 45° to 0° to − 45° to − 90°. With less vigorous specimens the responses becomes enfeebled under fatigue. I give below the results of a typical experiment carried out with a vigorous specimen, the response being distinguished as − when A is above, and + when A is below, the inversion bringing about a reversal direction of the responsive current.

Angle of inclination ...	+45°	+90°	+45°	0°	−45°	−90°
Galvanometer deflection ...	−19	−35	−18	0	+14	+25

RELATION BETWEEN ANGLE OF VERTICAL ROTATION AND INTENSITY OF GEOTROPIC REACTION.

The relation between the angle of inclination and the resulting geotropic action has already been determined by the Method of Axial Rotation. The ratio between the geotropic effects at 90° and 45° was thus found to be 1·49, which is nearly the same as $\frac{\text{Sin } 90°}{\text{Sin } 45°}$. I was next desirous of determining the relative excitations at the two angles by the Method of Vertical Rotation. It is necessary here to refer to certain differences of condition in the two methods. In the Axial Method, the hypothetical statoliths are distributed uniformly through the length of the cell, and rotation round the long axis causes displacement of the statoliths, the resulting pressure thus increasing with the sine of the angle of inclination. But in the case of vertical rotation through 45° to the right, the statoliths originally at the base of the cell accumulate to the right hand corner of the cell; a portion of the basal side of the cell is thus subjected to pressure. When the angle is increased to 90° the statoliths pass along the whole length including the basal and apical sides of the cell; but the excitability of the apical half may prove to be greater than that of the basal half. Hence excitatory geotropic effect is not likely to vary strictly as in sine of angle of inclination.

Whatever the reason may be, I find as a result of experiments with 12 different specimens that the mean ratio of the effects at 90° and 45°, obtained by the Method of Vertical Rotation, is 1·8 : 1 which is greater than 1·49 : 1 obtained by the Method of Axial Rotation, this latter value being practically the same as $\frac{\text{Sin } 90^\circ}{\text{Sin } 45^\circ}$.

SUMMARY.

It is shown that the state of excitation under direct stimulus is exhibited by an electrical change of galvanometric negativity; the effect of indirect stimulus induces, on the other hand, an electrical change of galvanometric positivity. The negative electric change corresponds to contraction and diminution of turgor; the positive electric change indicates, on the other hand, an expansion and increase of turgor.

The electric response to geotropic stimulus is studied by the two methods of Axial and Vertical Rotation. The upper side of a horizontally laid shoot is found to undergo an excitatory change of galvanometric negativity.

In quick reacting organs the latent period of geo-electric response is about 5 seconds, and the maximum excitation is induced in the course of 2 minutes.

The geo-electric response is due to physiological reaction. The intensity of response declines with age and is abolished at the death of the plant.

Under symmetrical conditions, the intensity of geotropic reaction is found proportional to the sine of the angle of inclination.

Electric investigation shows that the lower half of the pulvinus of *Mimosa* is geotropically more excitable than the upper half,

XLI.—THE MECHANICAL AND ELECTRICAL RESPONSE OF ROOT TO VARIOUS STIMULI

By

SIR J. C. BOSE.

IN the last chapter we studied the electric response of the shoot to the stimulus of gravity, and found that the excitatory effect of that stimulus is similar to that of other forms of stimulation. Before taking up the subject of the geo-electric response of the root to gravitational stimulus, I shall describe the effects of other forms of stimuli on the mechanical and electrical response of the root.

In connection with this subject, it should be borne in mind that the responsive curvature in the root takes place in the sub-apical growing zone which is separated by a certain distance from the tip. The stimulus is therefore direct when applied at the responding growing region; it is indirect when applied at the tip of the root. The intervening distance between the root-tip and the responsive zone of growth is semi-conducting or non-conducting.

I shall proceed to give an account of my investigations on the response of the root to direct and indirect unilateral stimulation. We shall study :—

(1) The Mechanical response to Direct unilateral stimulus.

(2) The Electrical response to Direct unilateral stimulus.

(3) The Mechanical response to Indirect unilateral stimulus.

(4) The Electrical response to Indirect unilateral stimulus.

MECHANICAL RESPONSE TO DIRECT STIMULUS.

As the geotropic responses of the shoot and the root are opposed to each other, the object of the investigation is to find out; whether the response of the root to various stimuli is specifically different from that of the shoot. We have seen that tissues in general respond to direct unilateral stimulus by contraction of the proximal and expansion of the distal side, the tropic curvature being thus *positive*. We shall now determine whether direct unilateral stimulation of the root induces a tropic movement which is similar or dissimilar to that exhibited by the shoot.

Experiment 178.—In experimenting with roots of various plants I obtained results which are precisely similar to that of the shoot. The movement of the root was observed by means of a reading microscope focussed on the tip of the organ. I employed various forms of stimuli, mechanical, thermal, and chemical. Unilateral application of these on one side of the growing region gave rise to a *positive* tropic curvature, resulting in a movement towards the stimulus. These experiments confirm Sachs' observation that unilateral application of stimulus in the region of growth induces positive curvature of the root.

ELECTRICAL RESPONSE TO DIRECT STIMULATION.

I next undertook an investigation on the electric response of the root to direct unilateral stimulation.

Experiment 179.—The terminals of the galvanometer were suitably connected with the two diametrically opposite points A and B in the growing region of the root. Stimulus was now applied very near the point A, the various stimuli employed in different experiments being: (1) mechanical, (2) chemical, and (3) thermal. In every instance the excited point A becomes galvanometrically negative. This shows that the response of the root is in no way different from that of the shoot.

MECHANICAL RESPONSE TO INDIRECT STIMULUS.

Before describing the effect of indirect stimulus on the root, I shall recapitulate its effects on ordinary tissues. I have shown that the effect of indirect unilateral stimulus is to induce a movement away from stimulus. This was shown to be the case with the bud of *Crinum* (p. 275) and the tendril of *Passiflora* (p. 291). The mechanical and electric response to indirect stimulation in the shoot is shown in the diagrammatic representation (Fig. 170). I shall now proceed to describe the mechanical response induced by unilateral stimulation of the root tip. As the responding region of growth is at some distance from the tip, the stimulation is therefore indirect.

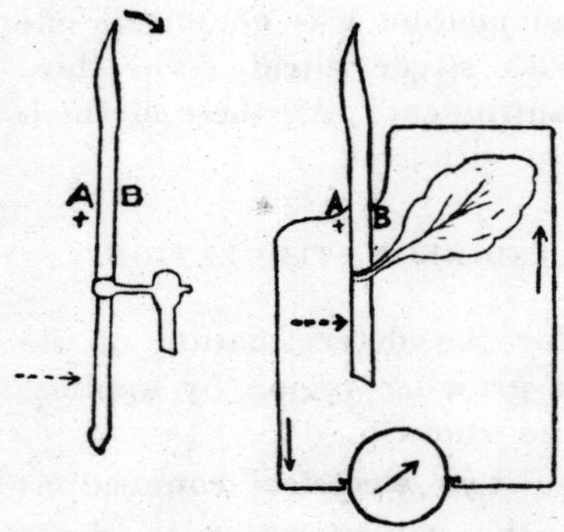

FIG. 170.—Mechanical and electrical response to indirect stimulation at dotted arrow. In figure to the left, the point A, on the same side undergoes expansion, with responsive mechanical movement away from stimulus indicated by continuous arrow. In figure to the right, indirect stimulus at dotted arrow induces electric response of galvanometric positivity at A, indicative of increase of turgor and expansion.

Experiment 180.—I employed at first mechanical stimulus of moderate intensity by rubbing one side of the tip of the root of *Bindweed*; this induced a movement away from stimulus. Unilateral application of dilute acid gave rise to a similar response. Thermal stimulus of moderate intensity also induced responsive movement away from the stimulus (Fig. 171).

Darwin in his *Movements of Plants* described experiments on the responsive behaviour of the tip of the radicle. He produced unilateral stimulation in three different ways, first by attaching minute fragments of cardboard to one side of the root-tip; this moderate and constant irritation was found to induce a convexity on the same side of the growing region, with the resulting negative movement, *i.e.*, away from stimulus. His second method was chemical, one side of the tip being touched with silver nitrate; the third method of stimulation was a slanting cut. All these methods induced a movement away from stimulus.

ELECTRICAL RESPONSE TO INDIRECT STIMULATION.

The next investigation was for the determination of the electrical change induced in the growing region by application of unilateral stimulus at the root-tip.

Experiment 181.—One of the two electrical connections with the galvanometer is made at one side of the growing region A, the other connection being made with the diametrically opposite point B. Unilateral stimulus was applied at the root tip *a*, of the bean plant and on the same side as A. I subjected the tip to various modes of unilateral stimulation. Mechanical stimulation was effected by emery-paper friction or by pin-prick; chemical stimulation was produced by application of dilute hydrochloric acid. Thermal stimulation was caused by the proximity of electrically heated platinum wire. In every case the response was by *induced galvanometric positivity at A* (Fig. 171). This

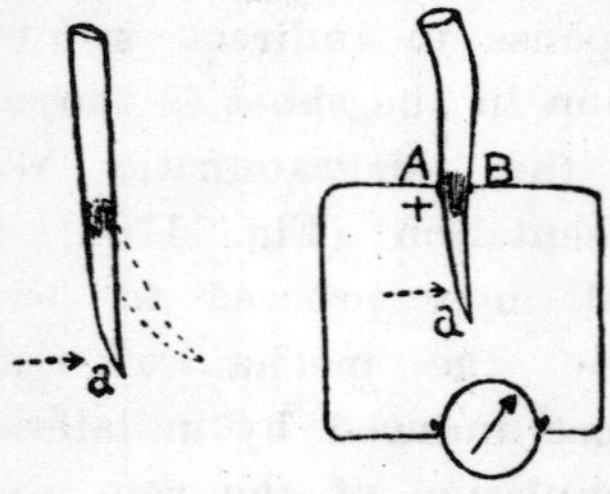

FIG. 171.—Diagrammatic representation of mechanical and electric response of root to indirect stimulus applied at the tip *a*. Figure to the left shows responsive movement away from stimulus. The electric response to indirect stimulus is indicated in the figure to the right; the point on the same side exhibiting galvanometric positivity. The shaded part indicates the responsive region of growth at some distance from the tip.

electrical variation took place within about ten seconds of the application of stimulus ; the interval would obviously depend on the length of path to be traversed by the transmitted effect of indirect stimulation.

The galvanometric positivity at A indicated that there was induced at that point an increase of turgor and expansion, in consequence of which the organ would move away from stimulus. Thus both by the mechanical and electrical methods of investigation we arrive at an identical conclusion that the effects of unilateral stimulus at the tip of the root gives rise to a movement, by which the organ is moved away from the source of stimulus ; since tropic movement towards stimulus is termed *positive*, this opposite response must be regarded as *negative*.

TABLE XXXVI.—EFFECT OF INDIRECT STIMULUS UNILATERALLY APPLIED AT THE ROOT-TIP.

Effect at the proximal side A in the growing region.	Effect at the distal side B.
Galvanometric positivity, indicative of increase of turgor and expansion.	Negligible.
The corresponding tropic curvature is negative, *i.e.*, a movement away from stimulus.	

The root-tip when burrowing its way underground comes in contact with hard substances and moves away from the source of irritation. The irritability of the root-tip is generally regarded as being specially evolved for the advantage of the plant. But reference to experiments that have been described shows that this reaction is not unique but exhibited by all plant organs, growing and non-growing. Indirect stimulus has been shown to give rise, in both shoot and root, to a *negative* tropic curvature in contrast to the *positive* curvature brought about by direct stimulation; the response of the root is therefore in no way different from that of vegetable tissues in general.

It will also be seen that an identical stimulus induces two opposite effects, according as the stimulus is applied at the tip or at the growing region itself. In the former case, the stimulus is indirect, and in the latter case it is direct. The results are in strict conformity with the laws of effects of direct and indirect stimulations that have been established regarding plant response in general (p. 231).

SUMMARY.

In the root, the responsive region is in the zone of growth. The tip of the root is separated from the region of response by a semi-conducting or non-conducting tissue.

Direct unilateral stimulus (applied at the region of growth) induces a positive curvature by the contraction of the proximal and expansion of the distal side.

The electrical response to direct unilateral stimulus is galvanometric negativity of the proximal, and galvanometric positivity of the distal side.

Indirect unilateral stimulus induces expansion of the proximal side resulting in negative curvature and movement away from stimulus.

The corresponding electric response induced is galvanometric positivity of the proximal side.

The responses of the root, to both direct and indirect stimulations, are precisely similar to those in the shoot. The assumption of specific irritability of the root as differing from that of the shoot, is without any justification.

XLII.—GEO-ELECTRIC RESPONSE OF ROOT

By

SIR J. C. BOSE,

Assisted by

SATYENDRA CHANDRA GUHA.

THE effects of various stimuli, direct and indirect, on the response of the root have been described in the last chapter. These responsive reactions have been found to be in no way different from those of the shoot. But the shoot and the root exhibit under the stimulus of gravity, responsive movements which are diametrically opposite to each other. These opposite effects of an identical stimulus have been regarded as due to specific differences of irritability in the two organs, specially evolved for the advantage of the plant. The root is thus supposed to be characterised by "positive" and the shoot by "negative" geotropism.

As regards response to other forms of stimuli, the root has been shown to behave like the shoot. We have now to inquire whether the reaction of the root to gravitational stimulus is specifically different to that of the shoot.

The electric method of investigation described in the last chapter, holds out the possibility of discovering the character of the responsive reaction induced in the root by its displacement from vertical to horizontal position; we shall, moreover, be able to make an electrical exploration of the root-tip and the zone of growth, and thus determine the qualitative changes of response, induced in two regions of the root under the action of gravitational stimulus. For the detection of geotropic action in the shoot, electric contacts were made at two points diametrically opposite to each other. Displacement of the shoot from vertical to horizontal position induced excitatory change of galvanometric negativity at the upper side of the organ, demonstrating the effect of direct stimulation of that side; this excitatory reaction of the upper side finds independent mechanical expression in the induced contraction and concavity of that side of the organ.

I employ a similar electric method for detection of geotropic excitation of the root, responses to geotropic stimulus being taken at the root-tip and also at the zone of growth in which geotropic curvature is effected. I shall now proceed to give a detailed description of the characteristic electric responses of the tip and of the growing region.

The two diametrically opposite contacts at the tip will be distinguished as *a* and *b*, the corresponding points higher up in the growing region being A and B. When the root is vertical the electric conditions of the two diametrically opposite points are practically the same. But when the root is rotated in a vertical plane through $+90°$ a geo-electric response will be found to take place; the direction of the responsive current disappears when the root is brought back to the vertical. Rotation through $-90°$ gives rise again to a responsive current, but its direction is found reversed.

GEO-ELECTRIC RESPONSE OF THE ROOT-TIP.

Experiment 182.—I took the root of the bean plant and made two electric contacts with the diametrically opposite points, *a* and *b*, of the root-tip at a distance of about 1·5 mm. from the extreme end. Owing to the very small size of the tip this is by no means an easy operation. Two platinum points tipped with kaolin paste are very carefully adjusted so as to make good electric contacts at the two opposite sides, without exerting undue pressure. For geotropic stimulation the root has to be laid horizontal, and as the root of the bean plant is somewhat long and limp, displacement from the vertical position is apt to cause a break of the electric contact. This is avoided by supporting the root from the top and also from the sides; for the latter purpose, I use paddings of cotton wool.

After due observance of these precautions the electric response obtained is found to be very definite; when the root is made horizontal, by rotation of the root through +90°, the point *a* is above, and the responsive current is found to flow from *b* to *a*, *the upper side of the tip* becoming galvanometrically negative; when the root is brought back to the vertical, the responsive current disappears; rotation through −90° makes the point *b* occupy

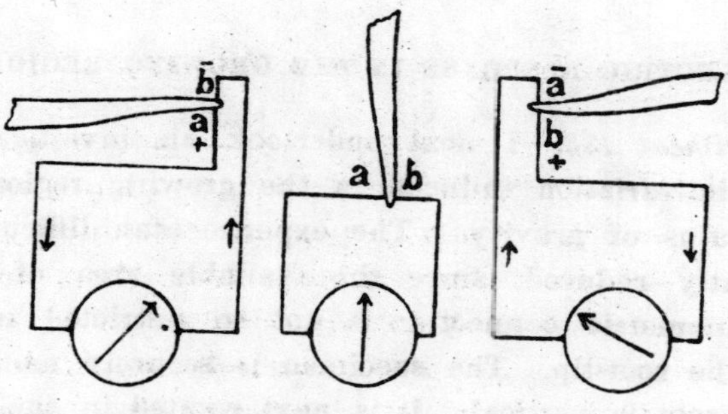

FIG. 172.—Diagrammatic representation of geo-electric response of root-tip. The middle figure shows root in vertical position. Rotation through + 90° places *a* above, which becomes galvanometrically negative. Rotation through − 90°, places *b* above and makes it negative.

the upper position, and the responsive current is from *a* to *b*; the upper side thus exhibits in every case, an excitatory electric change of galvanometric negativity (Fig. 172).

The root-tip thus exhibits the characteristic response to direct stimulation. Experiments carried out with 12 different specimens gave concordant results. The following table gives the absolute values of electro-motive force induced at the tip under geotropic stimulus.

TABLE XXXVII.—GEO-ELECTRIC RESPONSE OF THE ROOT TIP (*Vicia Faba*).

Specimen.	Induced E. M. F
1	0·0005 volt.
2	0·0011 "
3	0·0010 "
4	0·0015 "

ELECTRIC RESPONSE IN THE GROWING REGION.

Experiment 183.—I next undertook an investigation on the electric variation induced in the growing region under the stimulus of gravity. The experimental difficulties are here greatly reduced, since the available area of contact for galvanometric connection is not so restricted as in the case of the root-tip. The specimen is securely mounted so that the root is vertical. It is next rotated in the vertical plane through +90°, so that the point A in the growing region occupied the upper position. The electric response in the growing region took place in a short time and was very distinct. The induced electric change at A was now galvanometric *positivity* indicative of increase of *turgor and expansion.*

The series of experiments were carried out in the following order. The specimen was first rotated through +90° so that A was above. The responsive electric variation rendered it galvanometrically positive. The root was rotated back to neutral position when the current disappeared. The root was next rotated through −90° and the responsive current became reversed, the upper B becoming electro-positive (Fig. 173). The alternative rotations through +90° and −90° were carried out six times in succession with consistent results. The interval allowed between one stimulation and the next was determined by the period of complete recovery. Growing fatigue was

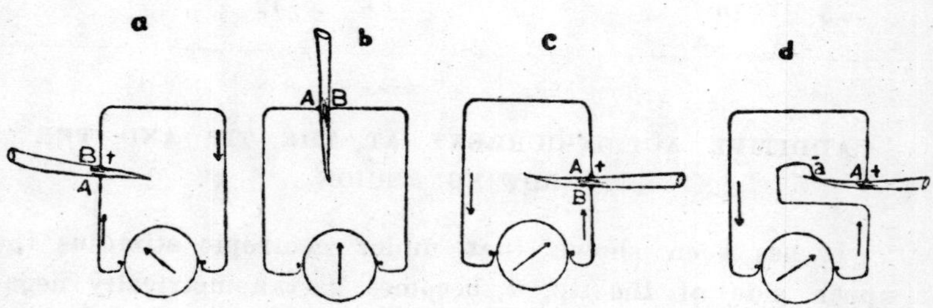

FIG. 173.—Diagrammatic representation of geo-electric response of growing region of root. (*a*) Rotation through −90° makes B, galvanometrically positive. (*b*) Vertical and neutral position. (*c*) Rotation through +90° places A above and renders it galvanometrically positive. (*d*) Additive effect on current of response, root-tip *a* negative, and growing region A positive.

found to increase this period; at first it was seven minutes, at the second repetition it was ten minutes, and at the third time it was prolonged to fifteen minutes.

I give below the series of electric responses induced by alternate rotations through +90° and −90°. The upper position was occupied by A in the odd series, and by B in the even series. In every case the upper side became galvanometrically positive.

TABLE XXXVIII.—GEO-ELECTRIC RESPONSE OF ROOT IN THE REGION OF GROWTH.

Odd series.	Galvanometer deflection A, positive.	Even series.	Galvanometer deflection B, positive.
1	20 divisions.	2	18 divisions.
3	16 ,,	4	18 ,,
5	10 ,,	6	12 ,,

ADDITIVE ACTION-CURRENT AT THE TIP AND THE GROWING REGION.

It has been shown that under geotropic stimulus the upper side of the tip, *a*, becomes galvanometrically negative, while the point A, higher up in the growing region, becomes galvanometrically positive. If now we make the two galvanometric connections with *a* and A, the induced electric difference is increased, and the galvanometric response becomes enhanced.

Experiment 184.—The root was at first held vertical, and two electric contacts made with *a* and A. In this neutral position there is little or no current. But as soon as the root was laid horizontal, an electro-motive response was obtained which showed that *a* was galvanometrically negative, and A galvanometrically positive (Fig. 173d). The induced electric response disappeared on restoration of the root to the vertical position. I give below the results of typical experiments with a vigorous specimen which gave strong electric response. It was possible to repeat the geotropic stimulation six times in succession, the results being perfectly consistent. The responses taken in succession exhibited slight fatigue, the first deflection being 140 divisions, and the sixth 115 divisions of the galvanometer scale.

TABLE XXXIX.—INDUCED E. M. F. VARIATION BETWEEN THE TIP AND THE GROWING REGION (*a* NEGATIVE AND A POSITIVE).

Geotropic stimulation.	Resulting electric response.
First stimulation ...	140 divisions.
Second " ...	130 "
Third " ...	130 "
Fourth " ...	123 "
Fifth " ...	127 "
Sixth " ...	115 "

The results of experiments 182 and 183 are summarised as follows:—

(1) the induced galvanometric negativity at root tip indicates direct stimulation of the tip, and

(2) the induced galvanometric positivity of the growing region shows that it is the effect of indirect stimulus that reaches it.

From these facts it will be seen that the tip perceives the stimulus and thus undergoes excitation, and that owing to the intervening tissue being a semi-conductor of excitation, it is the positive impulse that reaches the growing region and induces there an expansion and a convex curvature.

GEO-PERCEPTION AT THE ROOT TIP.

The results given above fully confirm Charles Darwin's discovery that it is the root tip that perceives the stimulus of gravity*; he found that removal of the tip abolished the geotropic response of the root. Objection has been raised about the shock-effect of operation itself being the cause of abolition of response. But subsequent observations have shown that Darwin's conclusions are in the main correct.

The experiments which I have described on the geo-electric response of the root tip and of the growing region offer convincing proof of the perception of the stimulus at the tip, and the transmission of the effect of indirect stimulus to the growing region. These experiments exhibit in an identical *uninjured* organ: the excitatory reaction at the upper side of the tip, the cessation of excitation, and the excitation of the opposite side of the tip, following the rotation of the organ through + 90°, 0° and − 90°. The effect at the growing zone is precisely the opposite to that at the tip, *i.e.*, an expansive

* "This view has been the subject of a considerable amount of controversy. Wiesner denies the localisation of geotropic sensitiveness. Czapek, on the other hand, supports Darwin's theory. Recently Picard has attacked the problem in a new way (and) concludes that not only the root tip but also the entire growing zone is capable of perceiving gravitational stimuli. . . . As both Picard's experimental method and his interpretation are open to criticism, the author has repeated his experiments with a more satisfactory apparatus. He finds that in *Vicia Faba*, *Phaseolus multeflorus* and *Lupinus albus*, both apex and growing zone are geotropically sensitive, the former being by far the more sensitive of the two, and the curvature of the growing zone being without a doubt largely induced by secondary stimuli transmitted from the apical region. Charles Darwin's views were therefore in the main correct."—Haberlandt—*Ibid*, p. 748.

reaction which results from the effect of indirect stimulus, in contrast to the contractile reaction due to direct stimulation.

We may now proceed a step further and try to obtain some idea of the difference in the mechanics of geotropic stimulation of the shoot and of the root, to account for the different responses in the two organs. The reason of this difference lies in the fact that in the shoot the perceptive and responding region is one and the same; every cut-piece of stem exhibits the characteristic geotropic curvature. In the root the case is different; for the removal of the sensitive root-tip reduces or abolishes the geotropic action; the region of maximum geotropic perception is thus separated from that of response. It must be borne in mind *that this holds good only in the case of gravitational stimulus*, for the decapitated root still continues to respond to other forms of stimulation such as chemical or photic.

The cause of this difference in the reactions to geotropic and other stimuli lies in the fact that in the latter case, energy is supplied from outside. But in geotropism the force of gravity is by itself inoperative; it is only through the weight of the cell contents that the stimulus becomes effective. Want of recognition of this fundamental difference has led many observers in their far-fetched and sweeping attempt, to establish an identity of reaction of the root to geotropic and photic stimulations, in spite of facts which plainly contradict it. Thus the root moves away from the incident vertical line of gravity; but under light, the root very often moves towards the stimulus. The negative phototropic response of the root of *Sinapis* is an exceptional phenomenon for which full explanation has been given in page 376.

We shall next consider whether the particular distribution of the falling starch-grains (which offers a rational explanation of geotropic stimulation) in the shoot and in the root, is capable of furnishing an explanation of the different geotropic responses in the two organs. In this connection, the results of investigation of Haberlandt and Nemec are highly suggestive. Haberlandt finds statoliths present in the responding region of the stem; the geotropic stimulation of the stem is therefore direct. Nemec's investigation on the distribution of statoliths in the root show, on the other hand, that it is the central portion of the root cap that contains the falling starch grains, and this would account for the indirect geotropic stimulation of the root.

The theory of statoliths is, however, not essential for the explanation of the opposite geotropic effects in the shoot and in the root. The observed fact, that the perceptive region in the root is separated from the responding region, is sufficient to explain the difference of geotropic action in the two organs. Through whatever means the stimulus of gravity may act, it is inevitable, from the fact that the stimulation of the shoot is direct and of the root indirect, that an identical stimulus should in two cases induce responsive reactions of opposite signs.

It will thus be seen that the postulation of two different irritabilities in the shoot and in the root is wholly unnecessary and unwarranted by facts. For the irritability of the root has been shown to be in no way different from that of other organs; an uniformity is thus found to exist in the reaction of all vegetable tissues.

SUMMARY.

On subjection of the tip of the root to the stimulus of gravity, the upper side exhibits excitatory reaction of galvanometric negativity. This shows that the root-tip undergoes direct stimulation.

The electric response in the growing region above the stimulated point of the root-tip is positive, indicative of increase of turgor and expansion. This is due to the effect of indirect stimulus.

The stimulus of gravity is perceived at the root-tip; it is the effect of indirect stimulus that is transmitted to the responding region of growth.

In contrast with the above is the fact that the growing region of the shoot is both sensitive and responsive to geotropic stimulus.

As the effects of direct and indirect stimulation on growth are antithetic, the responses of shoot and root to the direct and indirect stimulus must be of opposite signs.

There is no necessity for postulating two different irritabilities for the shoot and the root, since tissues in general exhibit positive or negative curvatures according as the stimulus is direct or indirect.

XLIII.—LOCALISATION OF GEO-PERCEPTIVE LAYER BY MEANS OF THE ELECTRIC PROBE

By

SIR J. C. BOSE,

Assisted by

SATYENDRA CHANDRA GUHA.

THE obscurities which surround the phenomenon of geotropism arise: (1) from the invisibility of the stimulating agent, (2) from want of definite knowledge as to whether the fundamental reaction is contractile or expansive, and (3) from the peculiar characteristic that the stimulus is only effective when the *external* force of gravity reacts *internally* through the mass of contents of the sensitive cells.

The experiments that have been detailed in the foregoing chapters will have removed most of the difficulties. But beyond these is the question of that power possessed by plants of *perceiving* geotropic stimulus by means of certain localised sense organs, which send out impulses in response to which neighbouring cells carry out the movement of orientation in a definite direction. Are the sensitive cells diffusely distributed in the organ or do they form a definite layer? Could we by the well established method of physiological response localise the sensitive cells in the interior of the organ? As the internal cells are not accessible, the problem would appear to be beyond the reach of experimental investigation.

It is true that post-mortem examination of sectioned tissues under the microscope enables us to form a probable hypothesis as regards the contents of certain cells causing geotropic irritation; we have thus the very illuminating theory of statoliths propounded by Noll, Haberlandt and Nemec. But for the clear understanding of the *physiological reaction* which induces the orientating movement, it is necessary to get hold, as it were, of a single or a group of sensory cells *in situ* and in a condition of fullest vital activity; to detect and follow by some subtle means the change induced in the perceptive organ and the irradiation of excitation to neighbouring cells, through the entire cycles of reaction, from the onset of geotropic stimulus to its cessation.

The idea of obtaining access to the unknown geo-perceptive cell in the interior of the organ for carrying out various physiological tests would appear to be very extravagant; yet I could not altogether give up the thought that the obscure problem of geotropic action might be attacked with some chance of success, by means of an electric probe which would explore the excitatory electric distribution in the interior of the organ. But the experimental difficulties which stood in the way were so great that for a long time I gave up any serious attempt to pursue the subject. And it is only when the present volume is going through the press that the very first experiments undertaken proved so highly successful that I am able to give a short account of the more important results, which cast a flood of light on the obscurities of geotropic phenomena. The new method has opened out, moreover, a very extensive range of investigation on the activities of cells in the interior of an organ, and enabled me to localise the conducting 'nerve' which transmits excitation in plants. These and other results will be given in the next volume.

METHOD OF EXPLORATION BY THE ELECTRIC PROBE.

The principle of the new method will be better understood if I first explained the steps of reasoning by which

I was led to discover it. The experiments described in Chapter XL showed that the upper surface of a horizontally laid shoot exhibits sign of excitation by induced galvanometric negativity; that this was due to the stimulus of gravity was made clear by restoration of the plant-organ to the vertical position, when all signs of electric excitation disappeared. Now the skin of the organ on which the electrode was applied could not be the perceptive organ, for the removal of the epidermis did not abolish the geotropic action; the perceptive layer must therefore lie somewhere in the interior. As every side of a radial organ undergoes geotropic excitation, the geo-perceptive cells must therefore be disposed in a cylindrical layer, at some unknown depth

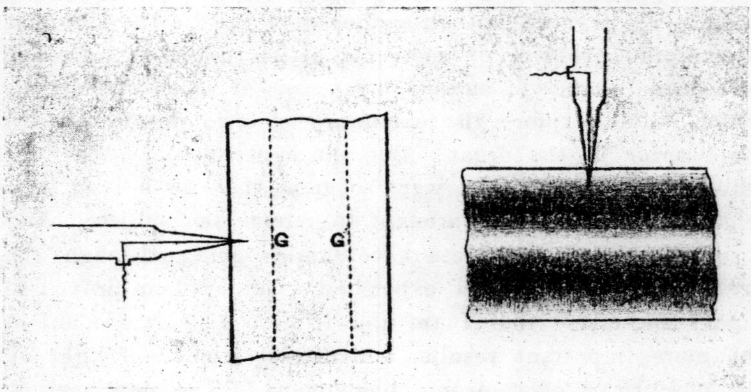

FIG. 174.—Diagrammatic representation of the geo-perceptive layer in unexcited vertical, and in excited horizontal position. (See text.).

from the surface. In a longitudinal section of the shoot, they would appear as two straight lines G and G′ (Fig. 174). In a vertical position the geo-perceptive layer will remain quiescent but rotation through + 90° would initiate the excitatory reaction. Let us first centre our attention to the geo-perceptive layer G, which occupies the upper position. This sensitive layer perceives the stimulus and is therefore the focus of irritation ; the state of excitation is, as we have seen, detected by induced galvanometric negativity, and the electric change would be most intense at the perceptive layer itself. As the power of transverse conduction is feeble, the excitation of the perceptive layer will irradiate into the neighbouring cells in radial directions with intensity diminishing with distance. Hence the intensity of responsive electric change will decline in both directions outwards and inwards.

The distribution of the excitatory change, initiated at the perceptive layer and irradiated in radial directions is represented by the depth of shading, the darkest shadow being on the perceptive layer. Had excitation been attended with change of light into shade, we would have witnessed the spectacle of a deep shadow (vanishing towards the edges) spreading over the different layers of cells during displacement of the organ from vertical to horizontal ; the shadow would have disappeared on the restoration of the organ to the vertical position.

Different shades of excitation in different layers is, however, capable of discrimination by means of an insulated electric probe, which is gradually pushed into the organ from outside. It will at first encounter increasing excitatory change during its approach to the perceptive layer where the irritation will be at its maximum. The indicating galvanometer in connection with the probe will thus indicate increasing galvanometric negativity, which will reach a maximum value at the moment of contact of the probe with the perceptive layer.

It will be understood that the surface electric reaction under geotropic stimulus, which we hitherto obtained, would be relatively feeble compared to the response obtained with direct contact with the maximally excited perceptive layer. When the probe passes beyond the perceptive layer the electric indication of excitation will undergo decline and final abolition. The characteristic effects described above are to be found only under the action of gravitational stimulus; they will be absent when the organ is held in a vertical position and thus freed from geotropic excitation.

I have hitherto spoken of the excitatory effect of the upper layer; there must be some physiological reaction on the lower perceptive layer, though of a different character, represented diagrammatically by vertical shading. Had the physiological reaction on the lower side of a radial organ been the same as on the upper, geotropic curvature would have been an impossibility, for similar reactions on opposite sides would, by their antagonistic effects, have neutralised each other.

After this preliminary explanation, I shall give a detailed account of the experiments and results. It is to be borne in mind that the investigation I am going to describe presupposes no hypothesis of geotropic action. I start with the observed fact that an organ under the stimulus of gravity, exhibits responsive movement. I ascertain the nature of the underlying reaction by electric tests; I have, in my previous works, fully demonstrated that the excitatory contractile reaction is detected by electro-motive change of galvanometric negativity, and the opposite expansive reaction by a change of galvanometric positivity. With the electric probe I ascertain whether geotropic irritation is diffuse, or whether it is localised at any particular depth of the organ. I map out the contour lines of physiological reaction with its heights and depths of excitation.

I shall now proceed to describe the results of electric exploration into the interior of the organ. The trouble I foresaw, related to the irritation caused by the passage of the probe, and the after-effect of wound on variation of excitability.

THE ELECTRIC PROBE.

The wound-irritation is, however, reduced to a minimum by making the probe exceedingly thin. A fine platinum

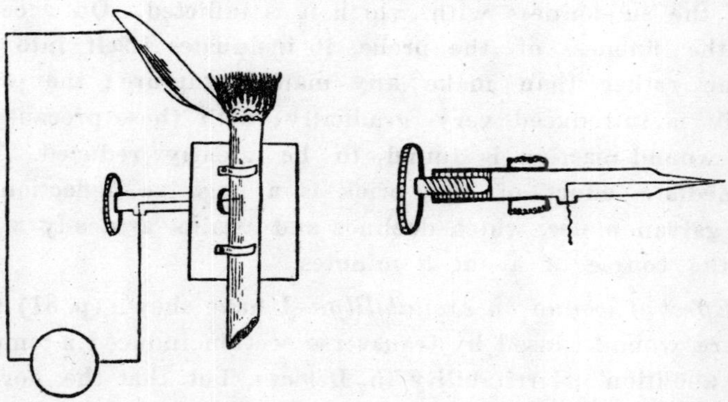

FIG. 175.—The Electric Probe. Figure to the left represents one electric contact made with sepal of *Nymphæa*, and the other, with the flower-stalk by means of the probe; the included galvanometer is represented by a circle. Figure to the right an enlarged view of the probe.

wire 0·06 mm. in diameter passes through a glass tubing drawn out into a fine capillary, and fused round one end of the platinum wire which protrudes very slightly beyond the point of fusion; the exploring electrode is thus insulated except at the protruded sharp point of the platinum wire. The length of the capillary is about 6 mm., just long enough to pass the experimental plant-organ transversely from one end to the other; the average diameter of the capillary is about 0·15 mm. The other end of the platinum wire comes out of the side of the tubing and is led to one terminal of the galvanometer, the other being connected with an indifferent point in the organ. The probe can be gradually pushed into the plant-organ by rotation of a screw head, one complete rotation causing a forward movement through 0·2 mm. (Fig. 175).

Wound-reaction.—I have shown that a prick acts as a mechanical stimulus, and in normal excitable tissues induces an excitatory change of galvanometric negativity. This wound-reaction increases with the extent of the wound, and the suddenness with which it is inflicted. On account of the fineness of the probe, it insinuates itself into the tissue rather than make any marked rupture; the probe again is introduced very gradually; with these precautions the wound-reaction is found to be greatly reduced. The immediate effect of the prick is a negative deflection of the galvanometer, which declines and attains a steady value in the course of about 5 minutes.

Effect of wound on excitability.—I have shewn (p. 81) that severe wound caused by transverse section induced a temporary abolition of irritability in *Mimosa*, but that the normal excitability was restored in the course of an hour. A prick from a thick pin was shown to depress temporarily the rate of growth, the normal rate being restored after an interval of 15 minutes (p. 202). In the case of geo-electric excitability, the depressing effect of the passage of the probe, I find, to disappear in the course of about 10 minutes.

For a choice of experimental material we have to find specimens which are not merely geotropically sensitive, but also exhibit large electric response under stimulus. In both these respects the shoot of *Bryophyllum* and the flower stalk of *Nymphæa* give good results.

ELECTRIC EXPLORATION FOR GEO-PERCEPTIVE LAYER BY MEANS OF THE PROBE.

Experiment 185.—I shall now proceed to give a detailed account of the experiments. The first specimen employed was the shoot of *Bryophyllum*, one contact being made with the side of the stem, and the other with an indifferent point on the leaf which was always held vertical. In a particular experiment, the probe was introduced into the stem through 0·4 mm. and a feeble galvanometric negativity was induced as the wound-effect. After an interval of 5 minutes, this attained a steady value of −15 divisions. On the rotation of stem through +90°, the point A was above and a very much larger deflection of −82 divisions was obtained, being the result of summation of wound and geo-electric effects. On restoration of the plant to vertical position the geo-electric reaction disappeared, leaving the persistent wound reaction of −15 divisions unchanged. The true geo-electric reaction at a point 0·4 mm. inside the stem was thus −67 divisions which is the difference between −82 and −15 divisions. I obtained in this manner the excitatory reactions at different layers of the organ. The following table gives true values of geo-electric reaction at different layers of the stem as the probe entered it by steps of 0·4 mm.

TABLE XL.—SHOWING THE GEO-ELECTRIC REACTION AT DIFFERENT DEPTHS OF THE ORGAN (*Bryophyllum*).

Position of the probe.		Geo-electric excitation (galvanometric negativity).
Surface	...	− 5 divisions.
0·4 mm.	...	− 20 "
0·8 "	...	− **24** "
1·2 "	...	− 22 "
1·6 "	...	− 18 "
2·0 "	...	− 14 "
2·4 "	...	− 10 "
2·8 "	...	− 5 "
3·2 "	...	− 0 "

The results given above, typical of many others, show that there is a definite layer in the tissue which undergoes maximum excitation under the stimulus of gravity, and that this excitation irradiates with diminishing intensity in radial directions inwards and outwards.

The geo-perceptive layer may thus be experimentally localised by measuring the depth of intrusion of the probe for maximum deflection of galvanometric negativity.

Localisation of geo-perceptive layer in **Nymphæa**: *Experiment 186.*—I employed the same method for the determination of the perceptive layer of a different organ namely, that of the flower stalk of *Nymphæa.* The electric reaction in *Nymphæa*, even under the prevailing unfavourable condition of the season, was moderately strong, being about three times greater than in *Bryophyllum.* A dozen observations made with different specimens gave very consistent results of which the following may be taken as typical. The probe was in this case, as in the last, moved by steps of 0·4 mm. at a time. Other examples will be given later where readings were taken for successive steps of 0·2 mm.

TABLE XLI.—SHOWING THE DISTRIBUTION OF INDUCED GEO-ELECTRIC EXCITATION IN DIFFERENT LAYERS (*Nymphæa*).

Position of probe.	Galvanometric deflection.
Surface ...	0 divisions.
0·4 mm. ...	− 16 "
0·8 " ...	− **42** "
1·2 " ...	− 20 "
1·6 " ...	− 10 "
2·0 " ...	− 2 "
2·4 " ...	0 "

It will be seen that as in *Bryophyllum*, so in *Nymphæa*, the geo-electric excitation increased at first with increasing depth of the tissue till at a depth of 0·8 mm. of the particular specimen the induced excitation attained a maximum value. The excitatory effect then declines till it vanished at a depth of 2·4 mm.

The depth of layer at which maximum excitation takes place varies to some extent, according to the thickness of the shoot. Thus while in a thin specimen of *Bryophyllum* 3·6 mm. in diameter the geo-perceptive layer was found at a depth of 0·6 mm., it occurred at the greater depth of 0·8 mm. in a thicker specimen, 5 mm. in diameter. In *Nymphæa* also the perceptive layer was found at a depth of 0·8 mm. in a thin and at a depth of 1·4 mm. in a thick specimen.

Having thus succeeded in localising the geo-perceptive layer by experimental means, it was now possible to examine the anatomical characteristics of the layer by examining it under the miscroscope. I also wished to find out from microscopic examination, the cause of certain differences noticed in the determinations of the perceptive layer in *Bryophyllum* and in *Nymphæa.* In the former the probe always encountered the maximally excited geo-perceptive layer from whichever point of the surface it entered the organ; this indicated that the sensitive layer in *Bryophyllum* was continuous round the axis. In *Nymphæa*, however, the probe occasionally missed the sensitive layer; but a new point of entry led to successful localisation of the perceptive layer; this was probably due to the particular layer not being continuous but interrupted by certain gaps.

MICROSCOPIC EXAMINATION OF THE MAXIMALLY EXCITED LAYER.

The specimens were taken out after the electric test, and the transverse sections made at the radial line of the passage of the probe. Thus in a particular experiment with *Bryophyllum* the point of maximum geotropic excitation was found to be at a distance of 0·8 mm. from the surface. By means of the micromoter slide in the stage and the micrometer eye-piece, the internal layer 0·8 mm. from the surface was examined; the particular sensitive layer S was recognised as the *continuous* 'starch sheath' or endodermis containing unusually large sized starch grains (Fig 176). These often occurred in loosely cohering groups of 8 to 10 particles, and their appearance is very different from the small sized irregularly distributed grains in other cells.

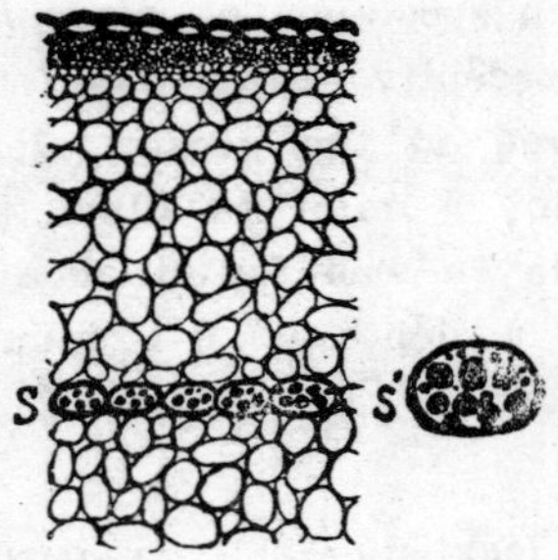

FIG. 176.—Transverse section showing continuous geo-perceptive layer S; enlarged view S' of cell of endodermis containing group of large starch grains. (*Bryophyllum*).

Examination of the microscopic section of the flower stalk of *Nymphæa* showed that the 'starch sheath' was not continuous but occurred in crescents above the vascular bundles which are separated from each other. The occasional failure of electric detection of the perceptive layer is thus due to the probe missing one of the crescents, which with intervening gaps, are arranged in a circle.

I give below a number of experimental determinations of the geo-perceptive layer in different specimens together with the micrometric measurement of the distance of the 'starch sheath' from the surface, the transverse section being made at the place where the probe entered the shoot.

Eight different determinations are given, three for *Bryophyllum* and five for *Nymphæa.*

TABLE XLII.—SHOWING THE POSITION OF THE GEO-PERCEPTIVE LAYER AND OF 'STARCH SHEATH' IN DIFFERENT SPECIMENS.

Specimen.	Distance of geo-perceptive layer from surface. (Method of electric probe.)	Distance of the starch sheath from surface. (Microscopic measurement.)
Bryophyllum :		
	(1) 0·6 mm. ...	0·6 mm.
	(2) 0·8 " ...	0·8 "
	(3) 0·8 " ...	0·8 "
Nymphæa :		
	(1) 0·6 " ...	0·6 "
	(2) 0·8 " ...	0·8 "
	(3) 0·8 " ...	0·8 "
	(4) 1·0 " ...	1·0 "
	(5) 1·4 " ...	1·4 "

Thus in all specimens examined, the experimentally determined geo-perceptive layer coincided with the 'starch sheath.' The theory of statoliths thus obtains strong support from an independent line of experimental investigation. The statolithic theory has been adversely criticised because in simpler organs the geotropic action takes place in the absence of statoliths. There is no doubt that the weight of the cell contents may in certain cases be effective in geotropic stimulation ; it may nevertheless be true that "at a higher level of adaptation, the geotropically sensitive members of the plant-body are furnished with special geotropic sense-organs—a striking instance of anatomico-physiological division of labour."*

In the instances of *Bryophyllum* and *Nymphæa* given above, the geo-perceptive layer localised by means of the

* Haberlandt—*Ibid*, p. 597.

electric probe is definitely found to be the endodermis containing large sized starch grains.

INFLUENCE OF SEASON ON GEO-ELECTRIC RESPONSE.

I shall now describe certain modifications in response, which result from the change of season and also from condition of high temperature. Physiological reactions, generally speaking, are much affected by different seasons; thus the seedlings of *Scirpus Kysoor* exhibit a very rapid rate of growth of 3 mm. per hour in August, but a month later the growth-rate declines to only 1 mm. per hour. I find similar depression of growth with the advance of season in seedlings of *Zea Mays*, where a very rapid fall in growth takes place in the course of a fortnight. The intensity of geotropic responses, both mechanical and electrical, of *Tropæolum* declines rapidly in the course of a month from February to March (p. 454). The flowers of *Nymphæa* began to appear by the end of June when the flower stalks exhibited strong geo-electric response. But later in the season, by July and the beginning of August, the response underwent continuous decline, and by the end of August the response was nearly abolished.

Much time had to be spent in perfecting the apparatus, and it was not till the beginning of August that the investigations could be properly started; the responsive indications were, however, marked and definite, though relatively feeble compared to those obtained at the beginning of the season. The decline of the geo-electric response was to a certain extent also due to the prevailing high temperature.

Effect of high temperature.—I shall in the next chapter describe experiments which show that geotropic response is diminished under rise of temperature. The specimens employed for localisation of geo-perceptive layer exhibited, as stated before, a decline of geo-electric response with the advance of the season. This may partly be due to unfavourable season, and partly to high temperature. In the middle of the season the responses were extremely feeble on warm days, but on cool mornings they became suddenly enhanced, to decline once more by the middle of the day. I could sometimes succeed in enhancing the sensitiveness by placing the specimen in a cold chamber. It thus appeared that certain internal change unfavourable for geo-perception takes place at high temperatures, and that the sensitive condition could sometimes be restored by artificial cooling. But later in the season, the internal change, whatever it may be, had proceeded too far, and artificial cooling did not restore the sensitiveness of the specimen. What are the physico-chemical concomitants which distinguish insensitive specimens, in which the electric indications had declined almost to the vanishing point?

TEST OF INSENSITIVE SPECIMENS.

I shall now describe the various physico-chemical concomitants which accompany the condition of relative insensibility. I have found three different tests: the electric, the geotropic, and the microscopic, by which the sensitive could be distinguished from the insensitive condition. The following tests were made on insensitive specimens.

Electric test: Experiment 187.—By the end of August the geo-electric indications given by the probe had, as stated before, almost disappeared. The tonic condition of the specimen, *below par*, was independently revealed by the response to prick of the probe; this, in vigorous specimens, is by an electric response of galvanometric negativity. But the response to prick in sub-tonic specimens is very different. I find that when the physiological condition of the tissue falls *below par*, the sign of response undergoes a reversal into one of *galvanometric positivity.* The same reversal under condition of sub-tonicity was also shown to take place in growth, where under the stimulus of light a positive acceleration took place, instead of normal retardation of growth (p. 221). In the present investigation, the insensitive specimens were found to give abnormal positive electric response to the stimulus of prick made by the probe. The prick-effect in fact often gave me previous indication as to the suitability of the particular specimen for exhibition of geo-electric response.

Test of geotropic reaction: Experiment 188.—I took four different specimens of *Bryophyllum* and *Nymphæa*, and held them horizontal. These plant organs had, earlier in the season, exhibited very strong geotropic effect, the shoot curving up through 90° in the course of ten hours or less. But these specimens obtained later in the season exhibited very feeble curvature, which hardly amounted to 10 degrees, even after prolonged exposure to geotropic action for 24 hours.

Test of microscopic examination.—I next made sections of *Bryophyllum* and *Nymphæa* and on examining them under the microscope discovered certain striking changes. A fortnight ago the group of large starch grains stained with iodine were the most striking feature of the starch sheath. But now these starch grains could not be found in any of the numerous specimens examined. The presence of the starch grains thus appears to be associated with the sensitiveness of the perceptive layer.

REACTION AT LOWER SIDE OF THE ORGAN.

There remains now the important question of the physiological change induced on the lower side of the horizontally laid shoot. The physiological reaction of two sides of the organ must be different, since the upper side exhibits contraction and the lower side expansion. It may be urged that the effect of one of the two sides might result from the passive yielding to the definite reaction induced on the opposite side. Investigation by the electric method enables us, however, to discriminate the two reactions from each other, since the electric response characteristic of the induced physiological change takes place in the organ, even under condition of restraint by which movement is prevented. We shall therefore investigate the geo-electrical reaction on the lower side of the securely held organ, and find out whether the induced electric change undergoes any variation in different layers from below upwards. There are two different ways in which the electric explorations of the lower side of the organ may be carried out. In the first method, the probe is introduced from below, and successive readings for geo-electric response taken as the probe enters the organ by successive steps. It is understood that the true geotropic effect is found from difference of galvanometer readings in vertical and horizontal positions. In the second method, the probe is introduced from above, and successive readings for the response taken for different positions of the probe as it enters the organ from the upper side and comes out ultimately at the lower side. This I shall call the METHOD OF TRANSVERSE PERFORATION. The intrusion of the probe on the upper side gives, as we have seen, increasing negative deflection of the galvanometer which reaches a maximum at the perceptive layer. Passage of the probe to still greater depths give deflections which decline to zero. But when the probe comes within the influence of the perceptive layer of the under side, the electric indication, as we shall presently find, undergoes a reversal.

ELECTRIC EXPLORATION OF THE LOWER SIDE OF THE ORGAN.

I shall first describe the results obtained from the first method, the probe entering the organ from the lower side.

Experiment 189.—The investigation was carried out with the stem of *Bryophyllum*, and the flower stalk of *Nymphæa.* The probe was made to enter the organ through 0·4 mm. and the geo-electric effect found, on rotation of the flower stalk of *Nymphæa* from the vertical to the horizontal. was a deflection of +6 divisions of the galvanometer. *The change induced at the lower side by geotropic stimulus is thus galvanometric positivity, indicative of enhancement of turgor, and of expansion.* Intrusion of the probe through 0·6 mm. gave rise to an increased positive geo-electric response. That the sign of electric response depended on the relation of the side of the organ to the vertical lines of gravity was demonstrated by alternate rotation of the plant through +90° and −90°, the probe remaining at a definite position. Rotation through +90 brought A above, and rotation through −90° brought A below. When the probe was in the *up* position the geo-electric response was negative, but when rotation through −90° brought it *below*, the response became positive. Thus with an identical contact in the plant, the electric response underwent reversal from negative to positive. This will be understood from the following table.

Position of the probe inside the organ.	Galvanometer deflection : A *above.*	Galvanometer deflection : A *below.*
0·4 mm. ...	− 8 divisions.	+ 5 divisions.
0·6 mm. ...	−16 „	+ 10 „

It will thus be seen that physiological change induced at any point is modified by its relation to vertical lines of gravity. When the point is above, the induced change is *negative*, when below, the induced change is *positive.*

I shall next describe the variation of effect at different layers of the under side of the organ.

Experiment 190.—A complete set of readings of the geo-electric reaction at different layers of the organ was taken, as the probe entered the lower side by successive steps of 0·2 mm. The following table gives the results obtained with a specimen of *Nymphæa.*

TABLE XLIII.—ELECTRIC EXPLORATION OF DIFFERENT LAYERS ON THE LOWER SIDE OF THE ORGAN (*Nymphæa*).

Position of the probe.	Galvanometer deflection.	Position of the probe.	Galvanometer deflection.
Surface ...	2 divisions.	1·0 mm. ...	**32** divisions.
0·2 mm. ...	4 ,,	1·2 ,, ...	16 ,,
0·4 ,, ...	8 ,,	1·4 ,, ...	12 ,,
0·6 ,, ...	16 ,,	1·6 ,, ...	4 ,,
0·8 ,, ...	20 ,,	1·8 ,, ...	0 ,,

It is thus seen that just as in the upper so also in the lower side, the electric variation undergoes at first an increase which attains a maximum; beyond this point the electric change undergoes a rapid decline. The induced electric change on the upper and lower sides are, however, different, galvanometric *negativity* in one case and *positivity* in the other.

The maximum galvanometric *negativity* of the upper side was found to occur at the geo-perceptive layer. We may next inquire about the anatomical characteristic of the layer in the lower side of the organ which exhibits the maximum galvanometric *positivity.* Microscopic section of the specimen employed in the above experiment showed the particular layer to be the starch crescent which lies above the vascular bundle. Thus the same geotropic layer which when placed above shows the maximum galvanometric negativity, exhibits maximum positivity when placed below

METHOD OF TRANSVERSE PERFORATION.

Experiment 191.—I next carried out a complete exploration of the interior of the organ along the diameter. The probe started from the upper surface, and came out at the lower by successive steps of 0·2 mm., the corresponding geo-electric effects being observed at each step. It has to be borne in mind that the successive readings were obtained by rotation from vertical to +90 (A above); the rotation was never carried out in the negative direction through −90°. But the probe entering from above passed the central axis, and entered a region where the galvanometric indication was transformed from negative to positive. The following table gives the results obtained with the flower stalk of *Nymphæa.*

TABLE XLIV.—SHOWING THE INDUCED GEO-ELECTRIC DISTRIBUTION ACROSS THE FLOWER STALK OF *Nymphæa* (diameter = 6·8 mm.)

Position of probe.	Galvanometer deflection.	Position of probe.	Galvanometer deflection.
Surface ...	− 10 divisions.	3·6 mm. ...	0 divisions.
0·2 mm. ...	− 26 ,,	3·8 ,, ...	0 ,,
0·4 ,, ...	− 40 ,,	4·0 ,, ...	0 ,,
0·6 ,, ...	− 50 ,,	4·2 ,, ...	+ 2 ,,
0·8 ,, ...	− 62 ,,	4·4 ,, ...	+ 4 ,,
1·0 ,, ...	− 72 ,,	4·6 ,, ...	+ 5 ,,
1·2 ,, ...	− 88 ,,	4·8 ,, ...	+11 ,,
1·4 ,, ...	− **108** ,,	5·0 ,, ...	+22 ,,
1·6 ,, ...	− 72 ,,	5·2 ,, ...	+38 ,,
1·8 ,, ...	− 44 ,,	5·4 ,, ...	+**46** ,,
2·0 ,, ...	− 30 ,,	5·6 ,, ...	+39 ,,
2·2 ,, ...	− 18 ,,	5·8 ,, ...	+32 ,,
2·4 ,, ...	− 10 ,,	6·0 ,, ...	+24 ,,
2·6 ,, ...	− 5 ,,	6·2 ,, ...	+18 ,,
2·8 ,, ...	− 2 ,,	6·4 ,, ...	+12 ,,
3·0 ,, ...	0 ,,	6·6 ,, ...	+ 6 ,,
3·2 ,, ...	0 ,,	6·8 ,, ...	+ 3 ,,
3·4 ,, ...	0 ,,		

A curve constructed from the data given above is seen in figure 177. The diameter of the flower stalk was 6·8 mm. The negative geo-electric reaction is seen to undergo an increase till it attains a climax at the depth of 1·4 mm. It then undergoes a continuous diminution till it becomes zero at the depth of 3 mm.; this neutral zone extends through 1 mm. When the probe enters a depth of 4·2 mm. measured from the upper side, it enters a region affected by the perceptive layer situated on the under side, the opposite physiological reaction being indicated by induced electric change of galvanometric positivity.

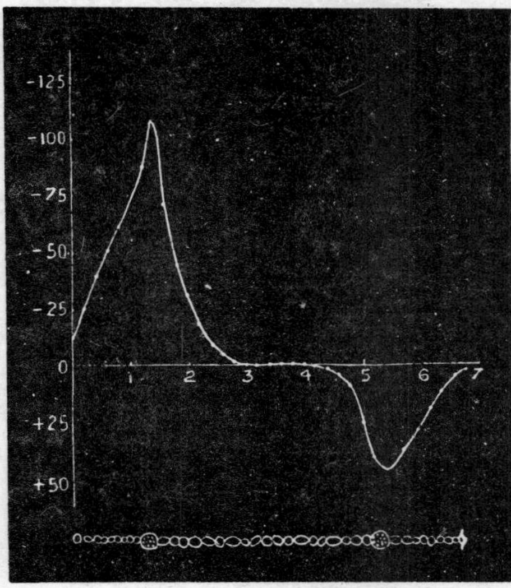

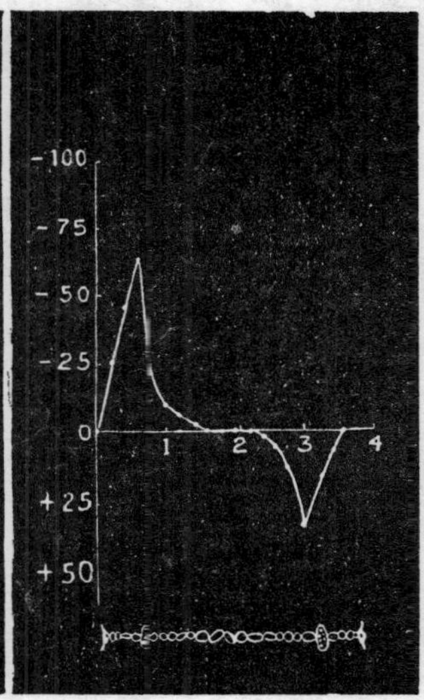

Fig. 177. Fig. 178.

FIG. 177.—Curve of geo-electric excitation in different layers of *Nymphæa*. Ordinate represents geo-electric excitation; abscissa, distance from upper surface of flower stalk. The diagrammatic section underneath shows the position of geo-perceptive layer (starch-sheath) corresponding to maximum induced galvanometric negativity and positivity on the two sides.

FIG. 178.—The curve of geo-electric excitation in different layers of *Bryophyllum*.

This positivity reaches a climax at a depth of 5·4 mm. measured from the upper side, and 1·4 mm. when measured from the lower side. The points of maximum positivity and negativity are situated symmetrically on the opposite sides of the organ. The electric variation of maximum positivity on the lower side is comparatively feeble, less than half the corresponding maximum negativity on the upper side. Microscopic section showed that the geo-perceptive layers were the same as the starch-crescents.

Experiment 192.—I carried out similar experiments with the shoot of *Bryophyllum.* The results are given in Table XLV; the curve of the electric distribution along the diameter is seen in figure 178. The characteristics of this curve are the same as that of *Nymphœa.* The maximum galvanometric negativity occurred at the depth of 0·6 mm., and of positivity at a corresponding point on the opposite side.

TABLE XLV. —SHOWING INDUCED GEO-ELECTRIC DISTRIBUTION ACROSS THE STEM OF *Bryophyllum* (diameter = 3·6 mm.).

Position of probe.	Galvanometric deflection.	Position of probe.	Galvanometric deflection.
Surface ...	0 divisions.	2·0 mm. ...	0 divisions.
0·2 m.m. ...	− 24 ,,	2·2 ,, ...	0 ,,
0·4 ,, ...	− 45 ,,	2·4 ,, ...	+ 3 ,,
0·6 ,, ...	− **63** ,,	2·6 ,, ...	+ 4 ,,
0·8 ,, ...	− 21 ,,	2·8 ,, ...	+ 9 ,,
1·0 ,, ...	− 9 ,,	3·0 ,, ...	+ **36** ,,
1·2 ,, ...	− 6 ,,	3·2 ,, ...	+ 21 ,,
1·4 ,, ...	− 3 ,,	3·4 ,, ...	+ 9 ,,
1·6 ,, ...	0 ,,	3·6 ,, ...	0 ,,
1·8 ,, ...	0 ,,		

Microscopic examination showed that the electric maxima in *Bryophyllum* coincided with the diametrically opposite points in the continuous endodermic ring. In *Bryophyllum* as in *Nymphœa*, the excitatory galvanometric negativity of the upper geo-perceptive layer is greater than the induced positivity of the lower layer in the ratio of about 2 : 1. But in a depressed condition of the tissue, the excitatory reaction is the first to disappear and the positive reaction persists, though with diminished intensity.

The geo-electric distribution in vigorous specimens seems to indicate that under the stimulus of gravity a marked excitatory reaction (contraction) takes place in the layer of cells contiguous to the upper geo-perceptive layer, and a less marked positive reaction (expansion) occurs in layers contiguous to the lower perceptive layer.

It is remarkable that physiological reaction of opposite kinds should occur on the upper and lower sides of an organ under the identical stimulus of gravity. The difference of reaction may conceivably be connected with the fact that the vertical lines of gravity enter by the upper, and leave by the lower side of the organ. The statolithic particles rest on the inner tangential walls of the perceptive cells of the upper layer, and on the outer tangential walls of the lower layer. Similar difference of physiological reactions of a polar character are also known in responses of plants under the action of an identical electric current; here with different ionic distributions, contraction takes place at the kathode, and expansion at the anode.

The geo-electric reactions that have been described were obtained under unfavourable conditions of climate and of temperature. But under better conditions the reaction becomes very greatly enhanced, as would appear from the following account of results which I obtained on two separate occasions in the beginning of August. The season had not become quite as unfavourable as towards the end of the month, but the prevailing sultry weather had caused great depression of the geo-electric excitability. On the first occasion referred to, thunderstorm had broken out at night, and it was refreshingly cool in the morning. It was with the utmost surprise that I noted the astonishing violence of the geo-electric response which the plants gave that morning; the maximum response hitherto obtained was about 100 divisions of the galvanomter scale; but on the present occasion the displacement of the plant, from vertical to horizontal position, induced responsive deflection so great that the galvanometer spot of light flew off the scale of 3,000 divisions. I was at first incredulous of the results and wasted the valuable occasion in trying to discover some hidden source of error. Subsequent tests showed that my misgivings were groundless, and that the extraordinary large deflection was really due to geo-electric reaction. On the second favourable occasion, which lasted for three hours (during the cool hours of the morning), I was able to secure a number of important observations. Thus displacement of the flower stalk of *Nymphœa* through $+90°$ was immediately followed by geo-electric response, the deflection being about 3,000 divisions of the scale. The latent period hardly exceeded a second; the return of the plant to the vertical position was quickly followed by electric recovery which was complete. The above results were obtained with the same specimen time after time without a single failure. The successive responses showed no sign of fatigue. Another remarkable effect was noticed during gradual increase of the angle of inclination. Nothing happened till a critical angle was reached, which was roughly estimated to be about 33°; when this critical angle was exceeded by a single degree, there was a sudden precipitation of geo-electric response. The experiments were repeated time after time with the identical result. It appeared as if some frictional resistance obstructed the displacement of the geotropic particles accumulated at the basal end of the cell, and it was not till the organ had been tilted beyond 33° that this resistance to sliding was overcome.

SUMMARY.

The electric distribution induced in an organ under the stimulus of gravity may be mapped out by means of an exploring Electric Probe.

The induced galvanometric negativity of the upper side of an organ (indicative of excitation) undergoes variation in different layers of the organ. The excitatory reaction attains a maximum value at a definite layer, beyond which there is a decline.

The geo-perceptive layer is experimentally localised by measuring the depth of intrusion of the probe for maximum deflection of galvanometric negativity.

The geo-perceptive layer thus determined is found to be the starch sheath which contains a number of large-sized starch grains.

The power of geo-perception undergoes seasonal variation. It is also lowered by high temperature.

The geo-electric response undergoes decline with growing sub-tonicity of the specimen; such specimens exhibit abnormal positive electric response under the stimulus of prick and feeble curvature under geotropic stimulus. The large-sized starch-grains, normally observed in the endodermis, are found to disappear in specimens which have become geo-electrically insensitive.

The electric response of the lower side of the organ to gravitational stimulus is of opposite sign to that of the upper side. The electric distribution on the lower side exhibits variations in different layers, the maximum positivity occurring at the perceptive layer. In vigorous specimens the excitatory negative electric change on the upper side is greater than the positive electric change on the lower side. Depressed condition of the tissue is attended by a relatively greater decline of the negative in comparison with the positive.

The induced electric variation on the upper and on the lower side indicates that the layers of tissue contiguous to the upper perceptive layer undergoes contraction, while those contiguous to the lower perceptive layer undergoes expansion.

XLIV.—ON GEOTROPIC TORSION

By

SIR J. C. BOSE,

Assisted by

GURUPRASANNA DAS.

I HAVE explained that in a dorsiventral organ, lateral application of various stimuli induces a responsive torsion by which the less excitable side is made to face the stimulus (p. 403). I shall in this chapter show that the effect of stimulus of gravity is in every respect similar to other forms of stimulation.

The direction of force of gravity is fixed, and we have to arrange matters in such a way that the geotropic sti-

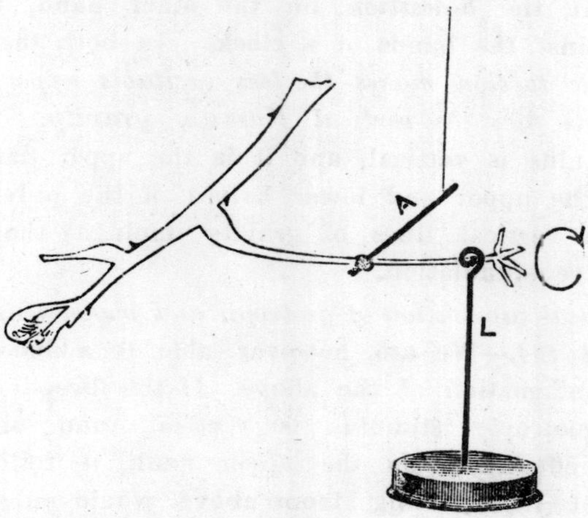

FIG. 179.—Diagram of arrangement for torsional response under geotropic stimulus. The less excitable upper half of pulvinus is, in the above figure, to the left and the torsional response is clockwise

mulus should act on the dorsiventral organ in a lateral direction. In the following experiments the pulvinus of

Mimosa is taken as the typical dorsiventral organ. For lateral stimulation, the plant is placed on its side, so that the vertical lines of gravity impinge on one of the two flanks of the organ. In regard to this, I shall distinguish two different positions, *a* and *b*. In the *a-position*, the apex of the stem and the upper half of the pulvinus are to the left of the observer, and in *b-position*, the apex of the stem and the less excitable upper half of the pulvinus are to the right. The arrangement for obtaining record of the torsional response under *a-position* is shown in figure 179.

Torsional response in a- and b-positions : Experiment 193.—When the leaf is in *a-position*, the geotropic torsion is found to be with the movement of the hands of a clock. In the *b-position*, on the other hand, the torsion is against the hands of a clock. In both these cases the *geotropic torsion makes the less excitable upper half of the pulvinus face the vertical lines of gravity.* The incident stimulus is vertical, and it is the upper flank, consisting of the upper and lower halves of the pulvinus (on which the vertical lines of gravity impinge) that undergoes effective stimulation.

Algebraical summation of geotropic and phototropic effects : Experiment 194.—We are, however, able to adduce further tests in confirmation of the above. If the direction of the incident geotropic stimulus is vertical, and should it act more effectively on the upper flank, it follows that stimulus of light acting from above would enhance the previous torsional response due to geotropism. In the above case, the lines of gravity and the rays of light coincide. The effect of rays of light acting from below should, on the other hand, oppose the geotropic torsion. The additive effect of stimulus of light and gravity is seen illustrated in figure 180. The first part of the curve is the record of pure geotropic torsional movement. Light from above is applied at L; the rate of movement is seen to become greatly enhanced. Light is next cut off, and the enhanced rate induced by it is also found to disappear, the response-curve being now due solely to geotropic action. The effect of geotropism in opposition to phototropism will be found in the following experiments, where the opposing action of light of different intensities is seen to give rise to a partial, to an exact, or to an over-balance.

BALANCE OF GEOTROPIC BY PHOTOTROPIC ACTION.

Photo-geotropic balance: Experiment 195.—I shall here describe in detail the procedure for obtaining an exact balance. A parallel beam of light from a small arc lamp is

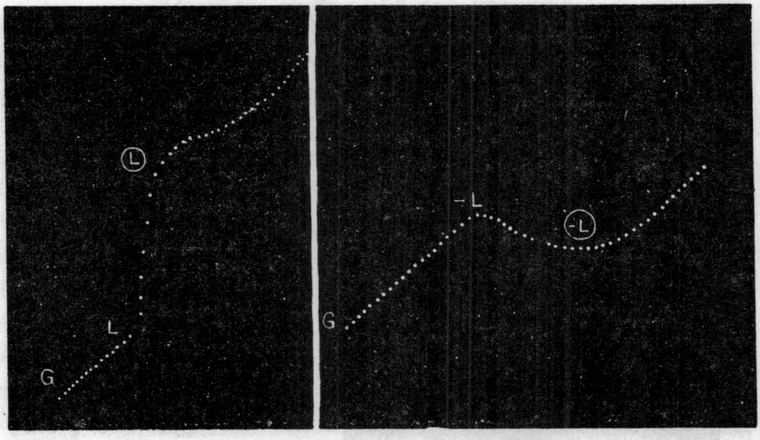

FIG. 180. FIG. 181.

FIG. 180.—Additive effect of stimulus of gravity G. and of light L. Application of light at—L increases torsional response. Removal of light restores original geotropic torsion.

FIG. 181.—Algebraical summation of geotropic and phototropic actions Light applied below at—L, opposes geotropic action. Cessation of light restores geotropic torsion. Cessation of light is indicated by L within a circle.

reflected by means of an inclined mirror, so as to act on the pulvinus below. An iris diaphragm regulates the intensity of incident light. The first part of the curve is the record of geotropic torsional movement. Light of a given intensity was applied below at a point marked – L (Fig. 181); this is seen to produce an over-balance, the phototropic effect being slightly in excess. The intensity of incident light was continuously diminished by regulation of the diaphragm till an exact balance was obtained as seen in the horizontal part of the record. It is with great surprise that one comes to realise the fact that the effect of one form of stimulus can be so exactly balanced by that of another, so entirely different, and that the stimulus of gravity could be measured, as it were, in candle powers of light! After securing the balance, light was cut off, and the geotropic torsion became renewed on the cessation of the counteracting phototropic action.

Comparative balancing effects of white and red lights: Experiment 196.—White light was at first applied at – L in opposition to geotropic movement. The intensity of light was stronger than what was necessary for exact balance, and its effect was at first to retard and then reverse the torsional response due to geotropism. When thus over-balanced, red glass was interposed on the path of light at R. As the phototropic effect of this light is feeble or absent, the geotropic torsion became predominant as seen in the subsequent up-curve. The red glass was next removed substituting white light at – L to act once more in opposition; the result is seen in the final over-balance, and reversal of torsion (Fig. 182).

FIG. 182.—Application of white light at—L in opposition causes reversal of torsion. Red light R, is ineffective, and geotropic torsion is restored. Reapplication of white light causes once more the reversal of torsion.

Effect of coal gas on the balance: Experiment 197.—The method of balance described above opens out new possibilities in regard to investigations on the relative modifications of geotropic and phototropic excitabilities by a given external change. Traces of coal gas are known to enhance the phototropic excitability of an organ while continued absence of oxygen is found to depress it. The experiment I am going to describe shows: (1) the enhancement of phototropic excitability on the introduction of coal gas, and (2) the depressing effect of excess of coal gas and of the absence of oxygen. After obtaining the normal curve of geotropic torsion, light was applied below at −L, and exact balance was obtained in the course of two minutes as seen in the top of the curve becoming horizontal. Coal gas was now introduced in the plant-chamber at C. This induced an enhancement of phototropic effect with resulting over-balance seen in the reversal of torsion. This enhancement persisted for more than three minutes. By this time the plant-chamber was completely filled with coal gas, and the resulting depression of phototropic action is seen in the second upset of the balance, this time in favour of geotropic torsion (Fig. 183). It would seem that the cells which respond to light are situated nearer the surface of the organ than those which react to geotropic stimulus. Hence an agent which acts on the organ

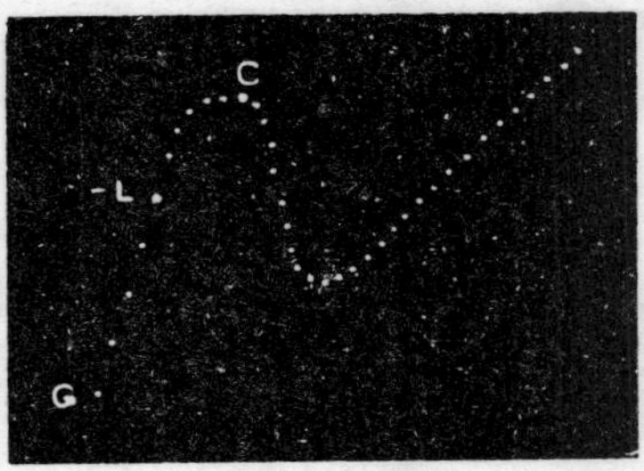

FIG. 183.—Effect of coal gas on photo-geotropic balance. Geotropic torsion, G, is exactly balanced by opposing action of light −L. Application of coal gas at C, at first caused enhancement of phototropic action with resulting reversal. Prolonged application induced depression of phototropic reaction, geotropic action thus becoming predominant.

from outside, induces phototropic change earlier than variation in geotropism.

SUMMARY.

Under lateral action of geotropic stimulus, a dorsiventral organ undergoes torsional response by which the less excitable half of the organ is made to face the stimulus.

The direction of incident geotropic stimulus is the same as the direction of vertical lines of gravity. Under geotropic stimulus it is the upper side of the organ that undergoes effective stimulation.

The effects of gravity and of light become algebraically summated under their simultaneous action. Light may be made to act in opposition to the stimulus of gravity. By suitable adjustment of the intensity of light, the two torsions become exactly balanced.

This state of balance is upset by any slight variation in one of the opposing stimuli.

The relative modification of geotropic and phototropic excitabilities by an external agent, is determined by the resulting upset of the photo-geotropic balance.

XLV.—ON THERMO-GEOTROPISM

By

SIR J. C. BOSE.

I SHALL in this chapter investigate the effect of variation of temperature on geotropic response. We have to bear in mind in this connection, that for the exhibition of geotropic curvature two conditions are necessary: (1) the presence of a perceptive organ to undergo excitation under the stimulus of gravity, and (2) the motility of the organ. A motile organ, including both the pulvinated and growing, will exhibit no geotropic effect on account of the depression of the power of perception through seasonal or other changes, or in the entire absence of the perceptive organ The organ may, on the other hand, possess the geo-perceptive apparatus, but no visible movement can take place in the absence of motility of the tissue.

As regards the modifying influence of temperature on geotropic curvature, the effect will depend on two factors:

(1) the influence of variation of temperature on geo-perception by the sensitive layer, and

(2) the modifying effect of temperature variation on the motile reaction.

I have in Chapter XLIII adduced facts which appear to show that the power of geo-perception declines at high temperatures. As regards motile reaction, we have seen that in *Mimosa* it increases from a minimum to an optimum temperature beyond which there is a depression (p. 55). As the optimum temperature for geo-perception is not necessarily the same as that for responsive curvature, the result is likely to be very complex.

The case becomes simpler after the attainment of maximum curvature. Enhanced temperature has a tendency to diminish the tropic curvature, as we found in the arrest and reversal of phototropic curvature under the application of warmth (p. 393); it appears as if rise of temperature induced a relatively greater expansion of the contracted side of the organ.

I shall now describe the effect of rising temperature on geotropic curvature in general, including torsion. A horizontally laid shoot curves upwards under geotropic action; a dorsiventral organ, owing to the differential excitabilities of its upper and lower sides, places itself in the so-called dia-geotropic position. A dorsiventral organ, moreover, exhibits a torsional movement under lateral stimulus of gravity.

In the geotropic movements we are able, as stated before, to distinguish three different phases (cf. Fig. 161). In the first, the movement initiated undergoes an increase; in the second, the rate of movement becomes more or less uniform; and in the last phase, a balance takes place between the tropic reaction, and the increasing resistance of the curved or twisted organ to further distortion.

The question now arises whether this position of geotropic equilibrium is permanent, or whether it undergoes modification in a definite way by variation of temperature. I shall proceed to show that the position of equilibrium undergoes a change in one direction by a rise, and in the opposite direction by a fall of temperature. I shall use the term *thermo-geotropism* as a convenient phrase to indicate the effect of temperature in modification of geotropic curvature and torsion.

I shall first deal with the effect of variation of temperature on geotropic torsion. Under the continued action of stimulus of gravity the torsion increases till it reaches a limit; for the twisted organ resists further distortion and a balance is struck when the twisting and untwisting forces are equal and opposite. In this state of equilibrium the effect of an external agent, say of variation of temperature, will bring about an upset of the balance. The torsion will be increased if the external agent induces an enhancement of geotropic action; it will, on the other hand, be decreased when it induces a diminished reaction.

A physical analogy will make this point clear; imagine a small magnetic needle suspended by a thin wire; the earth's directive force is supposed to be annulled by the well known device of a compensating magnet. A second and larger magnet M is now placed at right angles to the suspended needle; N will repel n and attract s, and a deflection will be produced, the deflecting force of the magnet M being balanced by the force of torsion of suspending wire (Fig. 184).

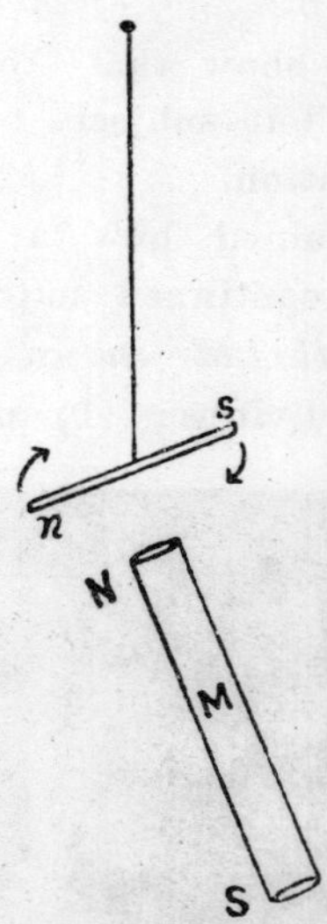

FIG. 184.—Magnet M causes deflection of the needle *n s*, suspended by a thin wire. Increase of magnetisation of M increases deflection, while decrease of magnetisation diminishes the deflection.

The state of equilibrium will however be disturbed by variation of the magnetic force of M. It is known that a rise of temperature diminishes magnetisation while lowering of temperature increases it. Hence the deflecting force of the magnet will be diminished under rise of temperature with concomitant diminution of deflection of the needle and the torsion of the wire. Fall of temperature, on the other hand, will cause an increase of deflection and of torsion. The physical illustration given above will help us to understand how the physiological effect of variation of temperature may bring about changes in geotropic curvature and torsion.

TROPIC EQUILIBRIUM UNDER VARYING INTENSITIES OF STIMULUS.

The following experiment will show that the position of tropic equilibrium is not fixed but subject to variation under changes of effective stimulation.

Experiment 198.—I have explained how a maximum tropic curvature is induced under continued action of light. Employing the pulvinus of *Erythrina indica* I applied light on the upper half of the pulvinus: (1) of medium

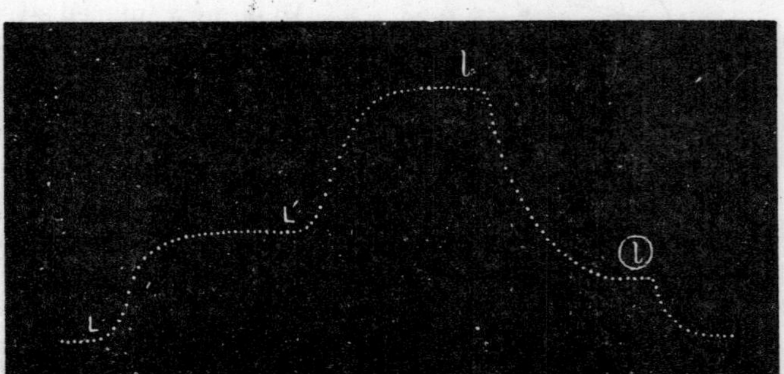

FIG. 185.—Effect of variation of intensity of light on phototropic equilibrium. Increase of intensity of light from L to L[1] produces an increased positive curvature and a new state of balance. Diminished intensity of light *l* brings about a new balance at a lower level. The cessation of light (*l* within a circle) restores the normal position of the organ.

intensity L, (2) of strong intensity L[1], and (3) of feeble intensity *l*. The source of light was an arc lamp; he intensity of light was varied by means of a focussing lens, which gave a parallel, a convergent or a divergent beam, with corresponding increase or diminution of intensity of light. Light was in each case continued till equilibrium was reached. Inspection of figure 185 shows that the position of equilibrium depends on the intensity of stimulation; the balance is 'raised' under increased and 'lowered' under decreased intensity.

In the case of geotropism the stimulus is constant, but its tropic effect, we shall presently see, undergoes variation with changing temperature.

EFFECT OF VARIATION OF TEMPERATURE ON GEOTROPIC TORSION.

Modification of geotropic torsion: Experiment 199.—The *Mimosa* plant was placed on its side, so that the pulvinus was subjected to lateral geotropic action. In response to this it underwent torsion, the upper half of the pulvinus tending to place itself so as to face the vertical lines of gravity. This torsional response was recorded as an up-movement; on the attainment of equilibrium the record became horizontal. The plant was now subjected to a cyclic variation of temperature, and the resulting variation of torsion recorded at the same time. The temperature of the plant chamber was gradually raised from the normal 30° to 34° C. and then allowed to return to the normal; finally the temperature was lowered to 26 C. Rise of temperature was effected by means of an electrical heater placed inside the chamber with a vessel of water placed above it. Care has to be taken that the rise of temperature is gradual, since a sudden variation often acts as a stimulus. The water in the vessel not only keeps the chamber in a humid condition but also prevents sudden fluctuation of temperature. After the temperature had been raised to 34 C., the heating current was stopped and the door of the plant chamber gradually opened, so as to allow the temperature to be restored to the normal. Cooled air was next introduced into the chamber till the temperature fell to 26°C. Figure 186 exhibits clearly the effect of variation of temperature on geotropic torsion. The maximum torsion had been attained at 30°C. and the first part of the record is therefore horizontal. Warmth was applied at H, and after a latent period of ten minutes, the geotropic torsion underwent a continuous diminution till a new state of equilibrium was reached at 34°C. This took place shortly after the stoppage of the heating current at (H). On return to normal temperature the torsional balance was restored to its original position of equilibrium. Application of cold at C, is seen to bring about a new state of balance with an increase of geotropic torsion.

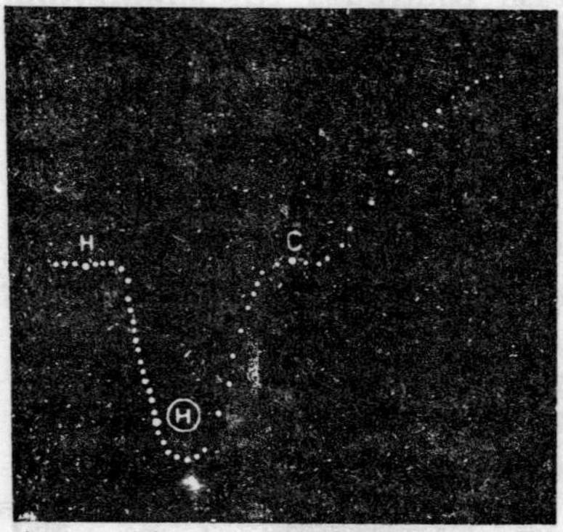

FIG. 186.—Effect of variation of temperature on geotropic torsion. Application of warmth at H diminishes the geotropic torsion; return to normal temperature (H) restores the original torsion; cooling at C, increases the geotropic torsion.

The position of geotropic equilibrium is thus seen to be modified by variation of temperature, the tropic effect being diminished with the rise, and enhanced with the fall of temperature.

It may be thought that the phenomenon just described may not be different from ordinary thermonasty, exhibited by the perianth leaves of *Crocus* and *Tulip* in which a rise of temperature induces a movement of unfolding, and a fall of temperature brings about the opposite movement of closure. In these cases the movement is determined solely by the natural anisotropy of the organ, and not by the paratonic action of a directive external force. Thus the inner side of the perianth leaves undergoes an expansion with rise of temperature attended by the opening of the flower; this movement of opening does not undergo any change on holding the flower in an inverted position.

But the torsional movement of the leaf of *Mimosa*, and the induced variation of torsion under change of temperature are not solely determined by the natural anistropy of the organ; it is, on the contrary, regulated by the directive action of the stimulus of gravity. The pulvinus in normal position does not exhibit any geotropic torsion and in the absence of an antecedent torsion change of temperature cannot induce any variation in it. It is only after the pulvinus had become torsioned under the lateral action of geotropic stimulus that a responsive variation is induced in it by the action of changing temperature.

The change in torsion is, moreover, determined in reference to the paratonic action of incident geotropic stimulus. This will be clearly understood from the tabular statement given below.

TABLE XLVI.—SHOWING THE EFFECT OF RISE OF TEMPERATURE ON GEOTROPIC TORSION.

Position of the organ.	Geotropic effect.	Effect of rise of temperature.
Right flank above : (*a*) position.	Right-handed torsion.	Left-handed torsional movement (untwist).
Left flank above : (*b*) position.	Left-handed torsion.	Right-handed torsional movement (untwist).

By right flank in the above table is meant the side of the pulvinus to the right of the observer facing the leaf of the plant held in the normal position. When the plant is laid on its left side in the *a*-position, the right flank will be above and the responsive torsion under geotropic stimulus becomes right handed or with the hands of a clock (Cf. Fig. 179). When the plant is laid on its right side, the left flank will be above and the geotropic torsion becomes left handed or against the hands of the clock.

It will be seen from the above that in whatever way the experimental condition may be varied, the movement in response to variation of temperature is determined in relation to the antecedent geotropic torsion. The geotropic effect whether left-handed or right-handed torsion is always diminished by the rise of temperature, and enhanced by the fall of temperature.

VARIATION OF APO-GEOTROPIC CURVATURE UNDER THERMAL CHANGE.

I shall now proceed to show that variation of temperature not merely induces variation of geotropic torsion but also of geotropic curvature. I shall first demonstrate the effect of thermal change on geotropic curvature of the shoot, and then demonstrate its effect on dia-geotropic curvature of leaves.

Experiment 200.—A specimen of *Tropæolum majus* grown in a small flower pot, is laid on its side. Under geotropic action the shoot becomes curved, the upper side becoming concave and the lower side convex. The end of the stem is attached to the recording apparatus; when the plant is subjected to a rise of temperature, the movement induced shows that the geotropic effect has undergone a diminution, the curvature exhibiting a flattening; lowering of temperature, on the other hand, increases the geotropic curvature. Other instances of this will be found in a subsequent chapter. The diurnal movement of the 'Praying Palm' is a striking example of the effect of variation of temperature in modification of geotropic curvature (p. 30).

Rise of temperature is thus shown to diminish geotropic torsion of dorsiventral organs, and the apo-geotropic curvature of radial organs. We have next to study the effect of temperature variation on the dia-geotropic equilibrium of leaves.

EFFECT OF VARIATION OF TEMPERATURE ON DIA-GEOTROPIC EQUILIBRIUM.

In the normal position of the plant, the leaf of *Mimosa* assumes, under geotropic action, an equilibrium position which is approximately horizontal. I shall proceed to show that this position of equilibrium also undergoes appropriate variation under changing temperature, the leaf undergoing a fall during rise, and an erection during fall of temperature.

I stated that the torsional response is one of the means of recording geotropic effect and its variations. In the ordinary position of the plant, the geotropic variation will be indicated by the responsive up or down movement of the leaf in a vertical plane. Taking the leaf of *Mimosa*, we have thus the means of studying the effect of variation of temperature by two independent means of inquiry, namely, by record of ordinary responsive movement in a vertical plane, and also by record of torsional response. The variation of temperature which induces these movements may be simultaneously recorded by means of a differential metallic thermometer. The Multiplex Recorder employed for this research consists of three recording levers. A photographic reproduction of the apparatus will be found in a subsequent chapter (see Fig. 190). The first lever is attached to the leaf of *Mimosa* placed in the normal position; the second lever records the torsional response of *Mimosa* leaf, the plant being placed on its side; the third lever attached to the differential metallic thermometer gives a continuous record of variation of temperature.

Effect of variation of temperature: Experiment 201.—Special arrangement was made for gradual variation of

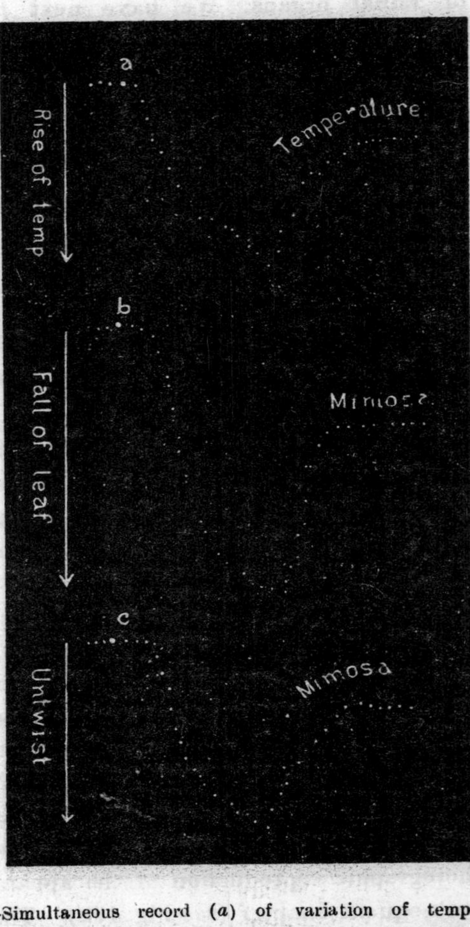

FIG. 187.—Simultaneous record (*a*) of variation of temperature, (*b*) of up or down movement of leaf of *Mimosa*, and (*c*) of variation of torsion. Rise of temperature is attended by fall of leaf and diminution of torsion, fall of temperature inducing the opposite effect.

temperature in the plant chamber. Two rectangular metallic vessels each 50 × 30 × 6 cm. were placed on opposite sides of the plant chamber, and warm water was made to circulate through them; this device ensured a steady rise of temperature. The flow of warm water was then stopped and the plant chamber was allowed to cool down; the fall of temperature was at first moderately rapid, but later on the rate of cooling became extremely slow; on account of this the temperature of the plant chamber, towards the end of the experiment remained higher than the normal temperature outside. The rate of rise and fall of temperature during the entire course is illustrated in the thermo-graphic (*a*) tracing (Fig. 187); the record (*b*) exhibits the movement of the leaf in a vertical plane, rise of temperature being attended by a diminution of geotropic curvature resulting in the fall of the leaf, the fall of temperature inducing the opposite effect. In record (*c*) is seen the responsive variation of geotropic torsion, rise of temperature inducing a diminution and fall of temperature causing an enhancement of torsion. The results obtained by diverse methods thus prove that the geotropic effect is diminished under rise, and increased under fall of temperature.

SUMMARY.

The position of equilibrium under geotropic action is not fixed but undergoes change with variation of temperature.

The geotropic curvature and torsion are increased by lowering of temperature, and decreased by rise of temperature. This is equally true of apo-geotropic and dia-geotropic curvatures.

PART IV.

NIGHT AND DAY MOVEMENTS IN PLANTS.

XLVI.—DIURNAL MOVEMENTS IN PLANTS

By

SIR J. C. BOSE.

THE subject has long been a perplexing one, and its literature is copious. After a good many years of experimental investigation, I have succeeded in analysing the main factors concerned in the many phenomena which have been described as Nyctitropism. The results of the researches are given in a sequence of five papers, which may be read separately, yet will be seen as so many chapters of what has been a single though varied investigation.

The different chapters are:

1. Daily movements in relation to Light and Darkness.
2. Daily movements due to Variation of Temperature affecting Growth.
3. Daily movements due to Variation of Temperature affecting Geotropic Curvature.
4. The Immediate and After-effect of Light.
5. Diurnal Movement of the leaf of *Mimosa* due to combined effects of various factors.

Nyctitropic movements are thus described by Jost*:

"Many plant organs, especially foliage and floral leaves take up, towards evening, positions other than those they occupy by day. Petals and perianth leaves, for example, bend outwards by day so as to open the flower, and inwards at night so as to close it . . . Many foliage leaves also may be said to exhibit opening and closing movements, not

* Jost—*Ibid*, p. 500.

merely when they open and close in the bud but also when arranged in pairs on an axis, they exhibit movements towards and away from each other. In other cases, speaking generally, we may employ the terms *night position* and *day position* for the closed and open conditions respectively. The night position may also be described as the *sleep position*." After reviewing the various theories proposed, he proceeds to say, "that a completely satisfactory theory of nyctitropic pulvinus movements is not yet forthcoming. Such a theory can only be established after new and exhaustive experimental research."

The difficulties of the experimental reinvestigation here called for towards clearing up and explanation of the subject are sufficiently great; they are further increased by the fact that these diurnal movements may be brought about by different agencies independent of each other. Thus in *Crocus* and in *Tulip*, the movement of opening during rise of temperature has been shown by Pfeffer to be due to differential growth in the inner and outer halves of the perianth. I shall in this connection show that a precisely opposite movement of closing is induced in *Nymphæa* under similar rise of temperature. I shall for convenience distinguish the differential growth under temperature variation as *Thermonasty* proper. Again certain leaflets open in light, and close in darkness in the so-called sleep position. Intense light, however, produces the 'midday sleep'—an effect which is apparently similar to that of darkness. The determining factor of these movements is the variation of light.

There are other instances of diurnal movement, far more numerous, which cannot be explained from considerations given above. It has therefore been suggested that the "Day and night positions may arise by the combined action of geotropism and heliotropism. Thus Vochting (1888) observed in the case of *Malva verticillatta*, that the leaves, when illuminated from below, turned their laminæ downwards during the day, but during the night became erect geotropically. The sleep movements in leaves and flowers, referred to above, cannot however be explained by assuming such a combination of heliotropism and geotropism."*

I commenced my investigation on nyctitropism five years ago, after having perfected an apparatus for continuous record of the movements of plants throughout day and night. A contrivance, described further on, has been devised for obtaining a record of diurnal variation of temperature. I have also succeeded recently, in perfecting a device for automatic record of variation of intensity of light. It has thus been possible not only to obtain a continuous record of the diurnal movement of the plant, but also obtain stimultaneous record of those changes in the environment which might have an influence on the daily movement. I have in this way collected several hundred autographs of different plants throughout all seasons of the year. The records thus obtained were extremely diverse, and it was at first impossible to discover any fundamental reaction which would explain the phenomenon. While in this perplexity my attention was directed two years ago to the extraordinary performances of the "Praying Palm" of Faridpore, in which the geotropic curvature of the tree underwent an accentuation during fall of temperature, and a diminution during rise of temperature.

The discovery of this new phenomenon led me to the inquiry whether Thermo-geotropic reaction, as I may call it, was exerted only on Palm trees, or whether it was a

* For further information on the subject of Nyctitropism, *cf.*—

Pfeffer—*Ibid*, Vol. II (1903), p. 112;

Jost—*Ibid*, pp. 500, 507;

Vines—Physiology of Plants (1886), pp. 406, 543.

phenomenon of universal occurrence. I therefore extended my investigation on various geotropically curved procumbent stems of *Ipœmia*, *Basella*, and of *Tropæolum majus*. Here also I found that diurnal variation of temperature induced a periodic movement exactly similar to that in Palm trees.

I next wished to find whether the Thermo-geotropic reaction observed in stems was also exhibited by lateral organs such as leaves, which being spread out in a horizontal direction are subjected to the stimulus of gravity. I found that in a large number of typical cases, a periodic movement took place which was exactly similar to that given by rigid trees and trailing stems. A standard curve was thus obtained which was found to be characteristic not only of trees and herbs, but also of leaves. The stem and leaves *fell* continuously with the rise of temperature, from the minimum at about 6 in the morning to the maximum at about 2 p.m. They erected themselves with falling temperature from 2 p.m. to 6 a.m. next morning.

In the diurnal record of *Mimosa* I met, however, with an unaccountable deviation from the standard curve, for which I could not for a long time find an adequate explanation. Subsequent investigations showed that the deviation was due to the introduction of additional factors of variation, namely of immediate and after-effects of light.

COMPLEXITY OF THE PROBLEM.

I have already referred to the great difficulty of explanation of nyctitropism from the fact that the diurnal movements may be brought about by different agencies independent of each other. It is, moreover, not easy to discriminate the effect of one agency from that of the other.

The combined effects of different factors will evidently be very numerous. This will be understood from consideration of the number of possible combinations with only two variables, geotropism and phototropism. The effect of geotropism may be strong G, or feeble, g. Similarly we may have strong effect of light L, or feeble effect of light l. Light may exert positive phototropic action $+L$ or negative action $-L$. Thus from two variables we obtain the following eight combinations:

$$G+L\ ;\ G-L\ ;\ G+l\ ;\ G-l\ ;$$

$$g+L\ ;\ g-L\ ;\ g+l\ ;\ g-l.$$

The number of possible variables are, however, far more numerous as will be seen from the following:

Geotropism.—The effect of geotropic stimulus on horizontally placed organs is one of erection. But this stimulus, which is constant, cannot by itself give rise to periodic movements. It has however been shown that variation of temperature has a modifying influence on geotropic curvature (p. 519).

Phototropism.—The action of unilateral light is to induce a tropic curvature, which in some cases is positive, in others negative (p. 386). In addition to these effects induced during the incidence of light, we have to take account of the after-effects on the cessation of light.

After-effects of light.—I find two very different effects, depending on the intensity and duration of previous illumination. Of these the most important is the phenomenon of 'overshooting' which occurs on the cessation of light of long duration. This particular reaction, to be fully described, will be found to offer an explanation of certain anomalous effects in diurnal movement.

Periodic variation of turgor.—I have shown (p. 39) that artificial enhancement of turgor in the plant induces an erectile movement of the leaf of *Mimosa*, diminution of turgor inducing the opposite movement of fall. Kraus and Millardet have shown that a diurnal variation of tension takes place in the shoot of all plants, which is presumably indicative of variation of turgor. This variation of turgor in the shoot must have some effect on the lateral leaves. But the leaves are subjected to conditions which are absent in the stem. The erect stem is, for example, free from geotropic action, whereas the lateral leaf is subject to it. The effect of turgor variation in the shoot on the movement of leaves may be, and often is, overpowered by the predominant geotropic action. I shall, later on, refer to this question in greater detail.

Autonomous movements : Experiment 202.—The lateral organ, say the leaf or leaflet, may have an autonomous movement of its own. In some, the autonomous movement may be relatively quick ; the complete pulsation in

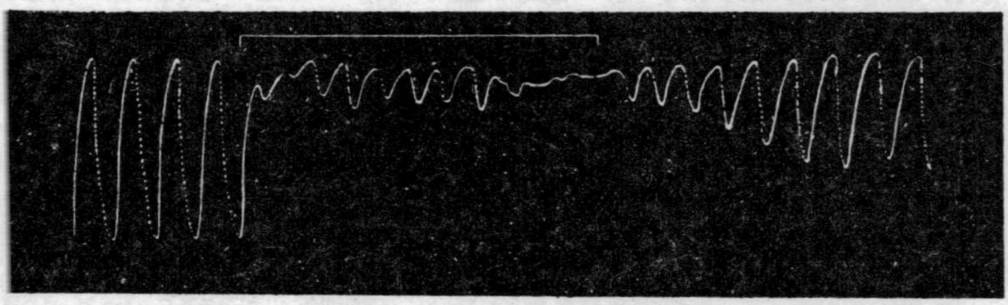

FIG. 188.—Arrest of pulsatory movement of leaflet of *Desmodium gyrans* by light from above and gradual restoration on cessation of light. Up-movement represented by up-curve.

Desmodium gyrans may be as short as a minute or so. I find that this autonomous movement becomes modified or even arrested by the paratonic effect of light. This is seen in figure 188, where light applied from above is seen to arrest the pulsation ; the normal activity is, however, restored on the stoppage of light.

Epinasty and Hyponasty : Experiment 203 **.—There are other autonomous movements which are relatively slow. Even in an erect stem there may be a to and fro oscillation. In such a case the effect of an external stimulus, say of light, is one of algebraical summation. The following is the summary of results of unilateral action of light on the nutating hypocotyl of a pea seedling :**

Natural movement.	Effect of light applied on the right side.
Movement to the right ...	Acceleration of existing movement.
Movement to the left ...	Retardation, arrest or reversal of natural movement.

Figure 189 exhibits the effect of light applied alternately above or below the cotyledon of *Cucurbita Pepo.* On

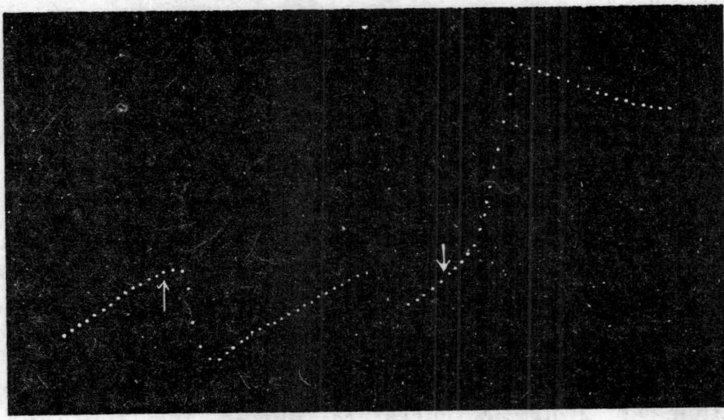

FIG. 189.—Effect of unilateral light on hyponastic movement of the cotyledon of *Pepo.* Application of light indicated by arrows; light acting from below retards, acting from above accelerates the movement. The last part of the curve in each shows recovery on the stoppage of light.

account of the more rapid growth of the lower side, the cotyledon was exhibiting a hyponastic up-movement. Application of light from above enhanced the existing rate

of movement, whereas light acting from below retarded the movement. Here we have instances of photo-hyponastic modification of natural movement. Similarly epinastic organs will, normally speaking, have their natural down movement retarded by light from above, and accelerated by light from below. If the periodicity of the autonomous movements coincides with the periodicity of the external stimulus, the resulting movement will be determined by algebraical summation ; it will be very pronounced when the two effects are concordant. If the two periodicities do not agree, the interference effects will become extremely complicated.

Positive thermonasty.—Rise of temperature inducing differential growth brings about the *closure* of the flower. Fall of temperature on the other hand induces a movement of opening. Example of this has already been given in the responsive movement of *Nymphœa.*

Negative thermonasty.—The opposite type of movement is exhibited by *Crocus* and *Tulip.* Pfeffer has shown that a rise of temperature induces in these flowers, a quicker rate of growth of the inner side of the perianth. Rise of temperature thus induces a movement of opening, and a fall of temperature brings about the opposite movement of closure. I shall presently describe the effects of both positive and negative thermonasty, in diurnal movements of flowers.

Thermo-geotropism.—I have already described the accentuation of geotropic curvature during the fall, and a flattening of curvature during the rise of temperature (p. 519). The influence of this factor on diurnal movement will presently be treated in fuller detail.

There are thus more than ten variables, and the resulting effect due to their combinations will exceed a thousand. This will explain why attempts at explanation of the phenomenon of nyctitropism had hitherto proved so baffling. It is indeed a difficult task to disentangle the full explanation of each given case in the vast complexity. It is, however, possible, by a process of judicious elimination, to reduce the difficulties which at first appear to be insurmountable.

In the periodic movement of plants there are several factors which are predominant, others being of minor importance. The important factors are the effects of light and darkness, of variation of temperature on differential growth, and of thermal variation on geotropic curvature.

For facility of treatment, I shall first take the three ideal types: (1) where the variation of light is the important factor, (2) where the movement is due to differential growth under variation of temperature, and (3) where thermal variation induces changes in geotropic curvature. I shall then take up the movement of the leaf of *Mimosa* where the combined effects of numerous factors give rise to a highly complex diurnal curve. There remains now the difficulty of discriminating the three types which approximate to the ideal.

DISCRIMINATING TESTS FOR CLASSIFICATION.

Predominant effect of light and darkness.—Turning first to the case where light exerts a predominant influence, the obvious test of keeping the plant in continuous darkness or continuous light is not practicable. One would think that if the movement was due to periodic variation of light, such movement would disappear under constant light or darkness. But owing to the persistence of after-effect, the periodic movement previously acquired is continued for a long time.

There is, however, another possibility of discrimination. The effect of variation of light will be most marked at the two periods, early in the morning when the light appears, and in the evening when it disappears. In the tropics there is little twilight; in Calcutta, the sun rises in summer at about 5-30 a.m., and sets at 6-30 p.m. In winter the sun rises an hour later, and the sunset is an hour earlier. The average dawn may therefore be taken approximately at 6 a.m., and the average sunset at 6 p.m. Unlike the diurnal variation of temperature which is gradual, the change from light to darkness or from darkness to light is very abrupt. If we succeed next in obtaining a continuous curve of the diurnal movement of the plant, the phototropic action would be evidenced by some flexures of the curve in the morning and towards evening.

The other two types of daily movement depend on the diurnal variation of temperature, and there is some difficulty in distinguishing the effect of variation of light from that of temperature, since both are connected with the appearance and disappearance of the sun.

Diurnal variation of light and of temperature.—There are certain differences, however, which enable us to distinguish the two variations. Light appears in the morning, say at 6 a.m., becomes most intense at noon; after 4 p.m. the light wanes, and darkness sets in quickly after 5 p.m. and remains persistent till next morning. The course of variation of temperature is somewhat different. The minimum temperature is attained in my green house at about 5 a.m. in summer, and at about 7 a.m. in winter. The maximum temperature is reached at about 3 p.m. in summer, and about 1 p.m. in winter. The range of daily variation in summer may be taken to be from about 23° C. to 34° C.; in winter it is from 16° C. to about 29°C. The above gives the normal variation and not the sudden fluctuations that occur during uncertain weather conditions.

The temperature remains constant for nearly an hour during the period of transition from falling to rising temperature, and *vice versâ.* The average period of minimum temperature may be taken at 6 a.m., which I shall distinguish as the *thermal-dawn.* The average period for maximum temperature, the *thermal-noon*, is at 2 p.m. Variatious from these average periods at different seasons do not amount to more than an hour.

The light-dawn and thermal-drawn are more or less coincident, while the thermal-noon is two hours later than the light-noon. A change in the diurnal curve of movement due to thermal variation will thus be detected at about 2 p.m. If the curve of daily movement of the plant-organ closely resemble the diurnal thermographic curve, there can then be no doubt of the causal relation of variation of temperature in the production of the periodic movement. Two different classes of phenomena, as already stated, arise however from the variation of temperature, *thermonasty* and *thermo-geotropism.* In the former, the movement is autonomous, and determined in relation to the plant; in the latter, the movement is related to the direction of external stimulus of gravity. Further tests will be given later, to distinguish the phenomenon of Thermonasty from that of Thermo-geotropism.

I shall in the succeeding papers describe the principal types of diurnal movements as sketched above. The success of the investigation greatly depends on the elaboration of automatic apparatus of precision, which gives a continuous record of the diurnal movement of different plant organs. The description of this Nyctitropic Recorder will be given in the next paper.

SUMMARY.

The obscurities in the nyctitropic movement of plants arise from the presence of numerous complicating factors.

In the diurnal movement of plants the most important factors are the effects of light and darkness, of variation of temperature on differential growth, and of thermal variation on geotropic curvature.

These three classes of phenomena may be discriminated from each other by the following tests. The effects of light and darkness are most pronounced in the morning when light appears, and in the evening when light disappears. A pronounced flexure in the diurnal curve at these periods indicates the dominant character of the phototropic action. The effect of light can also be distinguished from that of temperature from the fact that the period of maximum intensity of light, or *light-noon*, is about two hours earlier than the *thermal-noon*, at which the temperature is maximum.

A flexure of the diurnal curve about thermal noon, at which an inversion takes place from rise to fall of temperature, indicates the effect of temperature. The additional test of the effect of temperature is furnished by the close resemblance of the diurnal curve of the plant with the thermographic record for 24 hours.

Two different classes of phenomena arise from variation of temperature—Thermonasty and Thermo-geotropism. In the former the movement is autonomous and determined by the differential growth-activity of the two sides of an anisotropic organ. In the latter the movement is not in relation to the plant but directed by the external stimulus of gravity.

XLVII.—DIURNAL MOVEMENT DUE TO ALTERNATION OF LIGHT AND DARKNESS

By

SIR J. C. BOSE,

Assisted by

LALIT MOHAN MUKHERJI, B.Sc. (*Nawroji Scholar*).

THE nyctitropic movements of the leaflet of *Cassia alata* and of the terminal leaflet of *Desmodium gyrans* furnish us with typical examples of the recurrent effects of light and darkness. The petiole of *Cassia* contains a number of paired leaflets each of which is about 5 cm. long and 2·5 cm. broad. The leaflets are extremely sensitive to light ; at night each pair of leaflets fold themselves in a forward direction (see Fig. 150). With the appearance of light they open at first in a lateral direction ; later on there is a twist of the pulvinus by which the inner surface of the leaflets faces light coming from above (p. 405). I shall show that the diurnal movements of the leaflets are predominantly due to phototropic action.

Before proceeding further it will be necessary to give a general description of the experimental method employed, and of the apparatus by which diurnal movements are recorded.

EXPERIMENTAL ARRANGEMENTS.

The diurnal record is often taken continuously for several days, and it is therefore necessary to take precautions against the disturbing effect of watering the plant.

The record is also liable to be affected by the twist induced by light when it acts on one side of the organ.

Irrigation.—There is, as is well known, a periodic variation of turgor in the plant. This normal variation is, however, disturbed by watering the plant at irregular intervals. Precaution against this was taken by placing the three flower pots on a long trough filled with water (Fig. 190). The height of water in the trough is always maintained constant by a syphon.

Vertical illumination.—The direction of sunlight changes from morning to evening, and the leaves exhibit appropriate phototropic movements or torsions under changing directions of lateral light. In order to obviate this, a special chamber was constructed, which allowed light from the sky to fall vertically on the plant through a sheet of ground glass which covered the roof. The sides and the base of the chamber are impervious to light. A narrow slit covered with red glass allows inspection of the curve during the process of record.

The Ventilator.—A revolving ventilator, acted on by the wind, sucks the air away from the chamber, thus ensuring constant supply of fresh air, without causing any disturbances of the record.

The Recorder.—The Oscillating Recorder employed is of the quadruplex type carrying four recording plates (Fig. 190). The first lever records the daily variation of temperature. The other three are attached to three different specimens of the same plant, or to three different plants. In the former case, three records are obtained of the same species of plant, under identical external condition. If they agree in all essentials, the periodic curve may be taken as characteristic of the given plant. A very great saving of time is thus ensured, and it is thus possible to obtain characteristic curves of numbers of different species of

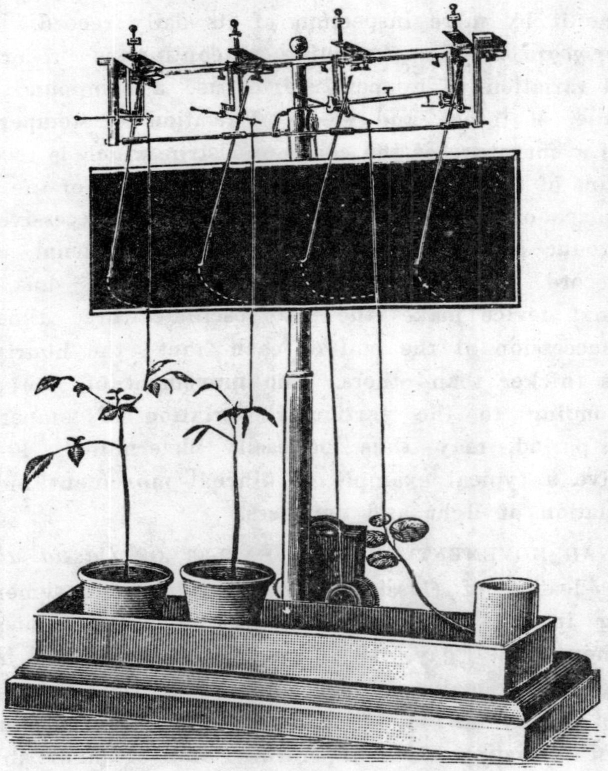

FIG. 190.—The Nyctitropic Recorder with four writing levers. The flower pots are placed in a trough filled with water to a constant height. The first two levers are shown in the figure to record movements of leaves, the third to record movement of a horizontally laid shoot; the fourth lever attached to a differential thermometer, T, records diurnal variation of temperature.

plants within the short period of a season. The quadruplex recorder enables us also to obtain simultaneous records under identical external condition of leaves of different age of the same plant, or of leaves of three different species of plant. I have for the last five years taken records of numerous plants at all seasons of the year. The autograph of the plant is often so characteristic that it is possible to name it by mere inspection of its daily record.

Thermograph.—For obtaining a continuous record of diurnal variation of temperature, I use a compound strip, T, made of brass and steel. Variation of temperature induces a curvature of the compound strip which is recorded by means of the attached lever. The oscillation of the plate takes place once in fifteen minutes, and the successive dots thus produced give time records of the diurnal curve. The record thus consists of a series of dots. An additional device makes the plate oscillate three times in rapid succession at the end of each hour; the hourly dot is thus thicker than others. The movement of the plant, corresponding to the particular variation of temperature at any period, may thus be easily determined. I shall now give a typical example of diurnal movement induced by variation of light and darkness.

DIURNAL MOVEMENT OF THE LEAFLET OF *Cassia alata.*

The leaflet of *Cassia alata* exhibits a movement of opening in the morning, and it remains outspread throughout the day. It then begins to close before evening and remains closed throughout the night. The problem before us is to find out the relative importance of variation of temperature and of light in the diurnal movement of the leaflets.

In the daytime the light is increasing till midday; there is, on the other hand, a rapid decline of light after 5 p.m. and uninterrupted darkness at night. As regards temperature there is a continuous rise from morning till the thermal noon at 2 p.m., after which the fall of temperature is continuous till next morning. The opening of the leaflets in the daytime may therefore be due to the summated effects of rising temperature and increasing light, the closure, on the other hand, being due to falling temperature, and to darkness. The individual effect of each of these factors is not known and it is therefore necessary to determine the effects of variation of temperature and of light.

EFFECT OF VARIATION OF TEMPERATURE.

Experiment 204.—The plant was enclosed in a glass chamber and exposed to diffuse light. The experiment was commenced at midday, when the leaflets were open; the light was kept uniform while temperature was artificially increased by means of an electric heater placed in the chamber, and decreased by introducing cold air into the plant chamber. One of the leaflets was attached to the recording lever and its movement, up or down, indicated the movement of opening or closure. The records showed that rise of temperature induces a movement of closure, while that of fall brings about the movement of opening.

EFFECT OF VARIATION OF LIGHT.

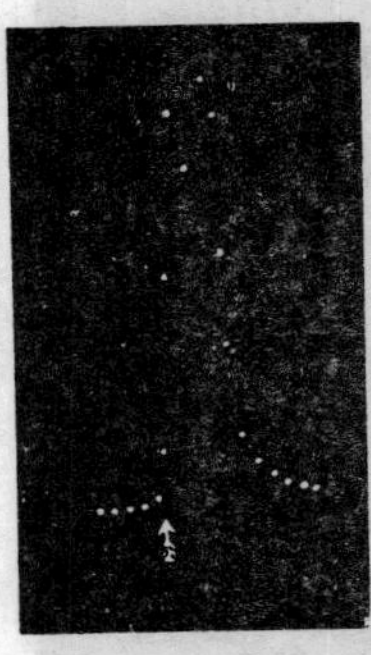

FIG. 191.—Effect of sudden darkening at arrow, produces movement of closure (up-curve). Restoration of light induces opening movement (down-curve). Successive dots at intervals of 15 minutes. (Leaflet of *Cassia*.)

Experiment 205.—This experiment was also carried out at midday, when the leaflets were open. The horizontal record in figure 191 represents the stationary expanded condition of the leaflet; a black cloth was put over the glass chamber at 1 p.m., and the effect of darkness was recorded for one hour. Darkness is seen to initiate a movement of closure which increased at a rapid rate; the black cloth was removed after an hour, and the movement of opening under light was completed in the course of five quarters of an hour. It is thus seen that the leaflets are extremely sensitive to the action of light.

The experiments that have just been described on the effects of rise of temperature, and of light, show that they are antagonistic to each other. In the forenoon the opening movement under light has to be carried out against the closure movement due to rise of temperature. Light, therefore, is the predominant factor in the diurnal movement of the leaflets of *Cassia*. The closure effect of darkness at night, on the other hand, overpowers the tendency of movement of opening due to fall of temperature.

DIURNAL MOVEMENT OF THE LEAFLET OF *Cassia alata*.

Experiment 206.—I next obtained the diurnal record of the leaflet, from 4 p.m. till 1 p.m. next day. The leaflets remain open from 1 p.m. to 4 p.m. and the record of this period is therefore omitted. In the diurnal record

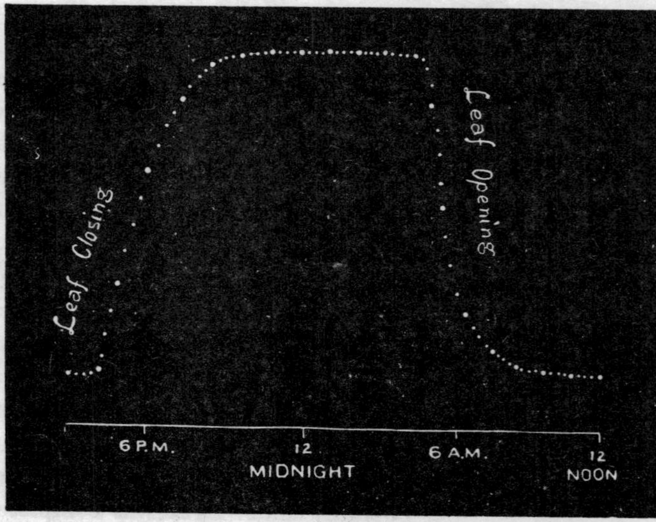

FIG. 192.—Diurnal movement of the leaflet of *Cassia alata*. Closure movement commenced at 5 p.m. and completed by 9 p.m. Leaflets began to open at 5 a.m.

(Fig. 192) the first thick dot was made at 4 p.m. and successive thick dots are at intervals of an hour, the thinner dots being at intervals of 15 minutes. It will be seen that a rapid movement of closure was initiated at 5 p.m. when the light is undergoing a rapid diminution. The movement of closure is completed at about 9 p.m. The leaflets remain closed till 5 a.m. next morning, after which they begin to open; this opening may commence even an hour earlier. It should be borne in mind in this connection, that since light and rise of temperature are antagonistic in their reactions, the effects of light and fall of temperature would be concordant; and the opening in the early hours may possibly be hastened by the low temperature in the morning. The leaflets open to their utmost by 9 a.m., and they remain open till the afternoon. The plant is so extremely sensitive to light that any slight fluctuation is followed by responsive movement of the leaflet. Thus the transitory passage of a cloud is marked in the record by a short-lived closure movement.

DIURNAL MOVEMENT OF THE TERMINAL LEAFLET OF *Desmodium gyrans*.

Both the petiole, and the terminal leaflet of this plant

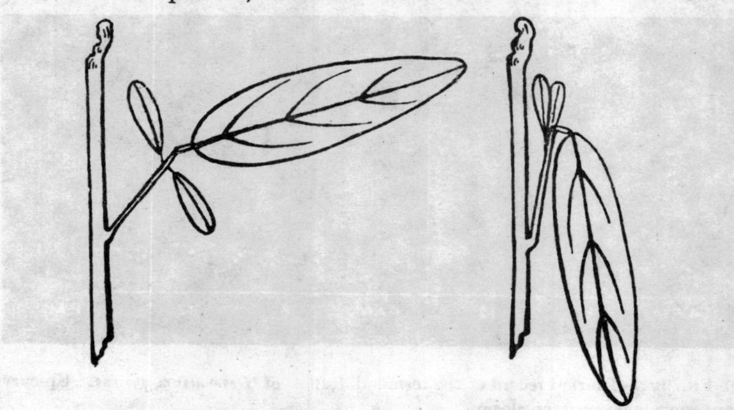

FIG. 193.—The day and night positions of the petiole and terminal leaflet of *Desmodium gyrans*.

exhibit very marked nyctitropic movement. The petiole is raised and becomes almost erect in the evening, while the pulvinus of the terminal leaflet exhibits a sharp curvature downwards (Fig. 193).

Experiment 207.—The petiole was held fixed, and the terminal leaflet attached to the recording lever. I have already explained that light falling on the pulvinus from above, induces an up-movement of the leaflet, which is thus erected under light of moderate intensity. If the light be strong, the transversely conducted excitation induces a partial neutralisation; very intense light may even cause a reversal into down-movement. Under natural conditions, day-light acting from above induces an up-movement; darkness, on the other hand, induces a rapid movement of fall. The leaflets sometimes exhibit autonomous pulsations; but the diurnal movement is very strong and the daily curve appears as a single large pulse on which smaller autonomous pulsations may become superposed.

The diurnal curve (Fig. 194) exhibits a sudden flexure at about 5 p.m. on the rapid waning of afternoon light till, by 6-30 p.m., it becomes closely pressed against the

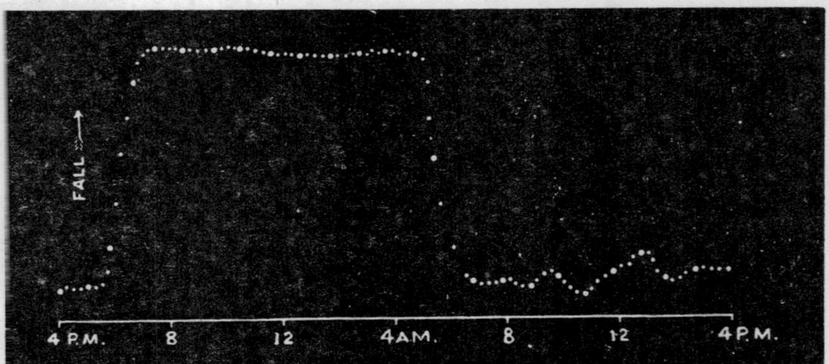

FIG. 194.—Diurnal record of the terminal leaflet of *Desmodium gyrans*. Up-curve represents movement of closure.

petiole, by the rapid fall of the leaflet. The discriminating test, between effects of variation of temperature and of light, lies in the fact that the flexure of the diurnal curve takes place in the former at about 2 p.m. when temperature undergoes change from ascent to descent; in the case of light, the change in the intensity of light begins to be marked about three hours later. In the diurnal curve of *Desmodium* the record shows little change at 2 p.m., showing that the leaflet is not affected to any great extent by the variation of temperature; it is, however, strongly affected by change in light as seen in the rapid closure movement about 5 p.m. The leaflet remains tightly closed throughout the night and begins to open and spread out early in the morning at about 5 a.m. This up-movement is also very rapid and the leaflet assumes the fullest outspread position by 7 a.m. It remains in this position till the afternoon, after which the cycle becomes repeated. As the leaflet is very sensitive to light, the position of equilibrium of the leaflet is liable to be disturbed by the slightest variation of light and the fluctuation of light from the sky often gives rise to a wavy outline in the record. The leaflet, moreover, has a tendency to exhibit rhythmic pulsations.

In the leaflets of *Cassia* and *Desmodium*, the daily movement is thus brought about by the predominant action of recurrent light and darkness.

MIDDAY SLEEP.

I shall here briefly recapitulate the results given in greater detail in an earlier paper (p. 352). I have shown that the midday closure of leaflets is brought about by the excitatory action of strong sunlight. The responsive movement of motile pulvinus under diffuse stimulus is determined by the greater contraction of the more excitable half of the organ. Under the action of the midday sun the leaflets of *Mimosa* undergo a folding upwards, whereas the leaflets of *Averrhoa carambola* a folding downwards. The explanation of the difference lies in the fact that in the leaflets of *Mimosa* it is the upper half, and in *Averrhoa* it is the lower half of the pulvinule, that is the more excitable. This difference may be demonstrated by the action of diffuse electric shock under which the leaflets of *Mimosa* exhibit an upward, and those of *Averrhoa* a downward, closure. I have also shown that conduction of excitation takes place across the pulvinule; hence the strong excitation caused by sunlight becomes internally diffused, and brings about the responsive movements, the direction of which is determined by the more excitable half of the pulvinule.

SUMMARY.

Rise of temperature induces a movement of closure of the leaflet of *Cassia*, fall of temperature inducing the opposite movement.

Artificial darkness induces a closure of the leaflets, the closure being completed in the course of an hour. On readmission of light, the leaflets become fully expanded in the course of one hour and a quarter. The leaflets are extremely sensitive to light, closure movement being induced by the transitory passage of a cloud.

The effect of rise of temperature is antagonistic to the action of light. The movement of opening during the course of the day is due to the effect of light overpowering the effect of rise of temperature.

Under daily variation of light and darkness, the movement of closure is initiated at about 5 p.m., when the light is undergoing a rapid diminution. The movement of closure is complete by 9 p.m. The leaflets remain closed till about 5 a.m. next morning, after which they begin to open and become fully expanded by 9 a.m.

The terminal leaflet of *Desmodium gyrans* exhibits a diurnal movement which is very similar to that of *Cassia*. It begins to open early in the morning and remains outspread during the whole day; the leaflet exhibits a rapid down-movement after 5 p.m. and becomes closely pressed against the petiole in the course of about two hours.

The midday sleep of leaflets of *Mimosa* and *Averrhoa* is due to the excitatory action of strong sunlight on the pulvinule, the more excitable half becoming contracted under excitation. In *Mimosa* leaflets it is the upper, and in *Averrhoa*, it is the lower half of the pulvinule that is the more excitable. It is in consequence of this that the diffuse excitation of strong sunlight causes the leaflets of *Mimosa* to fold upwards, those of *Averrhoa* to fold downwards.

XLVIII.—DIURNAL MOVEMENT DUE TO VARIATION OF TEMPERATURE AFFECTING GROWTH

By

SIR J. C. BOSE,

Assisted by

LALIT MOHAN MUKERJEE.

IT has been stated that there are two classes of diurnal movements caused by variation of temperature; one of these is due to differential growth induced on two sides of the organ, and the other is brought about by the induced variation of geotropic curvature. The former may be distinguished as *Thermonastic*, and the latter as *Thermo-geotropic movement.* Before laying down the criteria to distinguish the one class of phenomenon from the other, it would be advisable to refer to the somewhat arbitrary distinction that has been made between nastic and tropic reactions.

TROPIC AND NASTIC MOVEMENTS.

The explanation, which I shall offer about the night and day movements in plants, has been reached through the study not only of pulvinated, but also of growing and fully grown organs. A distinction is made between the movement due to growth, and the 'variation movement' due to change of turgor. I have shown (p. 239) that the same diminution of turgor which induces a contraction in a pulvinus, also induces in a growing organ an incipient contraction, and retardation of growth. Enhancement of turgor, on the other hand, induces in both the opposite effect of expansion. Unilateral stimulus induces curvature, and there is no essential difference in the production of such curvatures in pulvinated, growing, and fully grown organs. The exhibition of nyctitropic movement by the fully grown, and rigid 'Praying Palm' is a striking demonstration of the unity of response of all plant organs

As regards the distinction between the tropic and nastic movements, it will be found that there is no sharp line of demarcation between the two. A movement is said to be *tropic*, when unilateral stimulus acts on an organ and induces in it a directive movement. Curvature induced by diffused stimulus on a dorsiventral or anisotropic organ (with differential excitabilities of the two halves) is termed *nastic*. Daylight is supposed to act diffusely (*i.e.*, equally on all sides) on leaves; this is, however, not strictly true, since the light from sky above is stronger than from ground below. Moreover, the tropic action of unilateral light may become nastic by internal diffusion of excitation. This is seen in the response of the pulvinus of *Mimosa* to light acting from above. The leaf at first moves upwards towards the stimulus, the response being positively phototropic. But under the continued action of light, excitation becomes internally diffused, and the leaf undergoes a fall by the greater contraction of the more excitable lower half of the organ (p. 331). No sharp distinction can therefore be made between the movements of growth and of variation, between tropic and nastic curvatures.

The employment of the term 'nastic' is, however, convenient when used in a well-defined and restricted sense. "We speak of tropism when the organ takes up a resting position definitely *related to the effective stimulus*. Nastic movements, on the other hand, are curvatures which bring about a particular position *in relation to the plant*, and not to the direction of the stimulus".* It will sometimes be necessary, in the course of this paper, to discriminate

* Strasburger—"Text-book of Botany" (1912), p. 300.

the movements which are autonomous from others which are paratonic, *i.e.*, brought about by external stimulus, to the former class belongs a large number of automatic activities ranging from the quick pulsations of *Desmodium gyrans* to the slow movements, exhibited by epinastic and hyponastic organs. Under the category of nastic movements may also be included those of the flower of *Crocus* and *Tulip*, in which variation of temperature induces differential growth on two sides of the organ. The direction of the movement, though initiated by change of temperature, is determined by the difference of growth-activity on the two sides. In these instances of nastic movement, the induced curvature is in relation of the plant; the opening of the flower due to rise of temperature will remain the same, whether the flower be kept in an erect or in an inverted position. Had the movement, on the other hand, been paratonic, that is to say, due to the external stimulus of gravity, the responsive movement would have been determined not in relation to the plant but to the direction of external force of gravity.

In the description of direction of responsive movements, confusion is likely to arise unless the point of view be carefully defined. An up-movement of a leaf or a petal means approach towards the growing point of the axis. This may be variously described as movement of closure or of folding. A down-movement may, on the other hand, be described as a movement of opening or of unfolding. If the plant be held inverted, two different effects will be noticed depending on the character of the movement, whether nastic or tropic. In the case of nastic movement, the former up-movement in erect position would appear, on inversion of the plant, to be a down-movement; but in relation to the plant the closure movement will remain closure movement, whether the plant be held in the normal position or upside down. If, on the other hand, the direction of movement be determined by the paratonic effect of external stimulus, gravity for example, an up-movement due to fall of temperature will continue to be an up-movement, whether the plant be held in its normal or inverted position. The responsive movement in relation to the plant will, however, be different; the closure movement will, on inversion, be reversed into a movement of opening. The reversal of closure into an opening movement or *vice versâ* will thus be a test of the paratonic effect of external stimulus.

We may thus distinguish thermonastic from thermo-geotropic action by the following tests:

1. Thermonastic movements are, generally speaking, due to differential growth, and are therefore characteristically present in growing organs. Thermo-geotropic action is independent of growth.

2. Thermonastic movements take place in relation to the plant, and is not determined by external force of a directive nature. Opening or closing movement will remain unchanged after inversion of the plant. But thermo-geotropic reaction being determined by the external stimulus of gravity, becomes reversed on inversion of the plant. Closure movement is thus converted into opening movement, and *vice versâ*.

I shall now take up the diurnal movement due to variation of growth induced by change of temperature. Of this the flower of *Nymphæa* furnishes an example.

DIURNAL MOVEMENTS OF *Nymphæa*.

The flower of *Nymphæa* remains closed during the day and opens at night. Figures 195 and 196 are from photographs of the day and night positions of the flower. The closure and opening movements of this flower have

FIG. 195.—*Nymphæa* closed at daytime.

FIG. 196.—*Nymphæa* open at night.

been regarded as being mainly due to recurrent variations of light and darkness.* If the opening be due to darkness, closure of the flower should take place in the

* Pfeffer—*Ibid*, Vol. III, p. 122.

morning with the appearance of light. But the flowers often remain open till ten or eleven in the forenoon. I have sometimes succeeded in keeping the flower open for greater part of the day by lowering the temperature of the plant-chamber. The movement of the flower thus appeared to be associated with variation of temperature rather than of light.

Action of light: Experiment 208.—I investigated the effect of light on the movement of opening or of closing of the flower. One of the petals was attached to the recording lever; light from an arc lamp was made to act diffusely on the petal; this was done by means of two inclined mirrors by which the divergent horizontal beam of light was thrown on the upper and lower sides. The record in figure 197 shows that light induced a movement of opening, followed by closure in darkness. Since light induces a movement of opening, and darkness brings about a closure, the opening of the flower at night could not be due to darkness. We have therefore to look for a different cause for the diurnal movement of the flower.

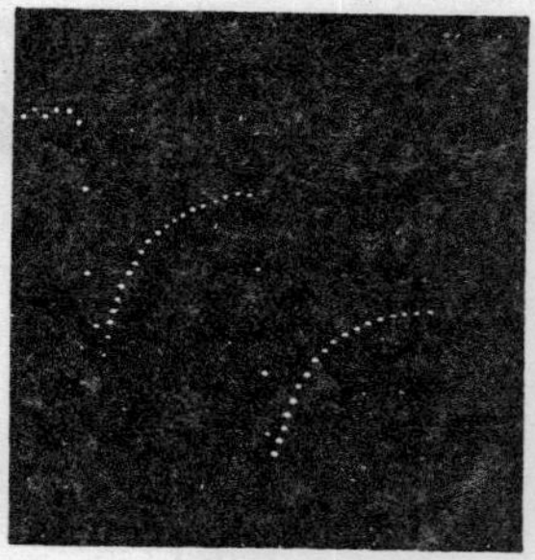

Fig. 197.—Response to light applied successively for 1 minute. Down-curve shows movement of opening followed by recovery in darkness. (*Nymphæa*).

Effect of variation of temperature.—I have already described an experiment which proves that rise of temperature induces a movement of closure of the floral leaves of *Nymphæa*, lowering of temperature producing the opposite effect (p. 311).

From the study of the action of light and of variation of temperature, it will be seen that the flower of

Nymphœa is acted on in the evening by two antagonistic forces; darkness induces a movement of closure, and fall of temperature gives rise to a movement of opening. Since the flower opens in the evening, the predominant effect is that of falling temperature.

The above conclusions are fully borne out by the diurnal record which I obtained with *Nymphœa.*

Experiment 209.—One of the perianth leaves was attached to one of the recording levers, the differential thermometer being attached to the other. It will be seen (Fig. 198) that the movement of the flower follows very

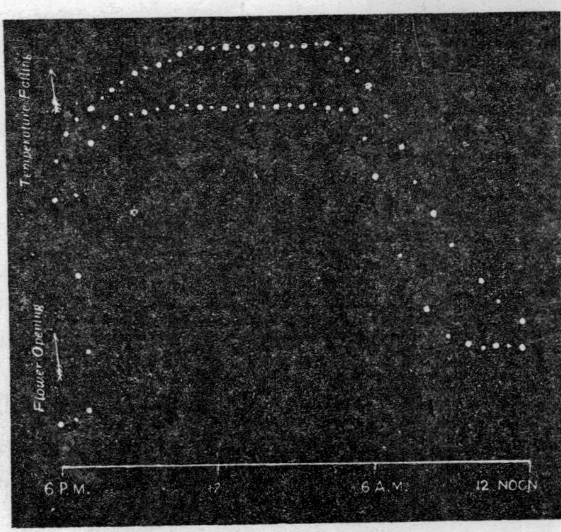

Fig. 198.—Diurnal record of *Nymphœa.* Upper record gives variation of temperature; the up-curve representing fall, and down-curve rise of temperature. The lower record exhibits the movement of the flower, up-curve representing the opening, and down-curve the closure of the flower.

closely the curve of variation of temperature. The flower was tightly closed in the day time; and the perianth leaves began to open out in the evening at first slowly, then very rapidly, and the flower becoming fully expanded by 10 p.m.

at night. Though the temperature continued to fall, there was no possibility of further expansion beyond the maximum. The temperature began to rise after passing through the minimum at 6 a.m., and the movement of closure set in with rising temperature, the flower becoming completely closed by 10 a.m. That geotropism has little effect is seen from the fact that the inversion of flower does not interfere with the normal opening or closing of the flower.

The phenomenon of diurnal movement of *Nymphæa* is therefore thermonastic, the floral leaves exhibiting movement of opening at night owing to fall of temperature. *Luffa acutangula*, which opens in the afternoon, and closes early in the morning, gives a diurnal record similar to that of *Nymphæa*.

SUMMARY.

The flower of *Nymphæa* exhibits a movement of closure during rise of temperature, and of opening during fall of temperature.

It is shown further that the effects of light and of rise of temperature are antagonistic to each other. Light is shown to induce in *Nymphæa* the movement of opening, and darkness to cause the movement of closure. The diurnal movement of *Nymphæa* is not therefore due to periodic variation of light and darkness, but to the predominant effect of variation of temperature.

The diurnal record shows that the perianth leaves begin to open in the evening with falling temperature, and the flower becomes fully expanded by 10 p.m. The movement of closure sets in with rising temperature in the morning, and the flower becomes fully closed by 10 a.m.

XLIX.—DAILY MOVEMENT IN PLANTS DUE TO THERMO-GEOTROPISM

By

SIR J. C. BOSE,

Assisted by

LALIT MOHAN MUKHERJI.

OF the vast number of daily movements perhaps the largest proportion is due to thermo-geotropic reaction and its modifications. Thermo-geotropic movements have the following characteristics:

1. The organs are sensitive to the stimulus of gravity and the periodic movements are brought about by variation of geotropic curvature under change of temperature.

2. The movement is not confined to growing organs, but is also exhibited by organs which are fully grown and even by rigid trees.

3. The periodic movement is closely related to the diurnal variation of temperature. Fall of temperature from thermal-noon (about 2 p.m.) to thermal-dawn (about 6 a.m.) is attended by a movement of erection; rise of temperature from thermal-dawn to thermal-noon is followed, on the other hand, by a reverse movement of fall.

That the movement is primarily due to variation of temperature will be demonstrated in two different ways:

(*a*) by the change of normal rhythm of movement by artificial transpositions of periods of maximum and minimum temperature, and

(*b*) by the abolition of periodic movement through maintenance of constant temperature.

That the phenomenon is not nastic, but paratonic will be demonstrated :—

(*a*) by the reversal of closure into opening movement and *vice versâ*, in consequence of inversion of the plant upside down, and

(*b*) by the diurnal variation of torsional movement, the direction of which is dependent on the directive action of the stimulus of gravity.

I shall now describe the diurnal movement of various geotropically curved plant-organs; the most striking example of this is furnished by the 'Praying' Palm of Faridpore, already described. I shall here recapitulate some of the important features connected with the phenomenon.

DIURNAL MOVEMENT OF PALM TREES.

Movements similar to that of the Faridpore Palm (p. 12) are found in other Palm trees growing at an inclination from the vertical. I reproduce once more the diurnal curve given by the Sijberia Palm together with the curve of daily thermal variation (Fig. 199). It will be seen that the two curves resemble each other so closely that the curve of movement of the tree is practically a replica of the thermographic record. There can therefore be no doubt of the movement being brought about by variation of temperature; rise of temperature is attended by the movement of fall of the tree and *vice versâ*. The record was commenced at noon; the temperature rose till the maximum was reached at about 3 p.m. and the tree also reached its lowest position at 3-45 p.m., the lag being 45 minutes. The temperature fell continuously after the maximum at 3 p.m., to the minimum at 6 a.m. next morning. In response to the falling temperature, the tree exhibited a movement of erection. The temperature rose after 6 a.m. and the movement of the tree became reversed from ascent to descent.

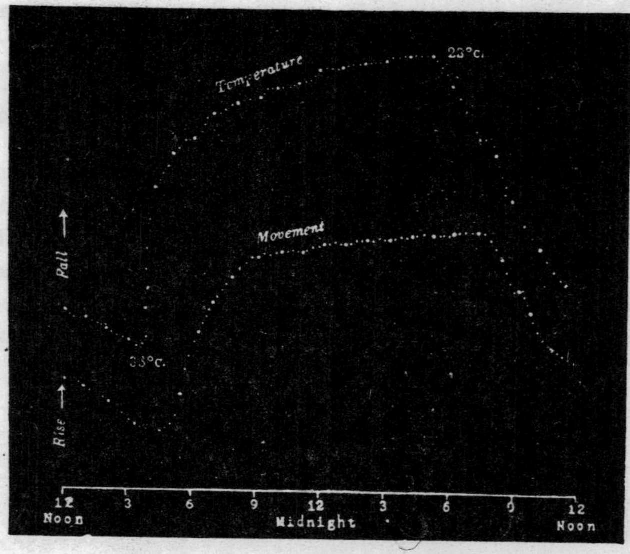

FIG. 199.—Diurnal record of the Sijberia Palm. Upper curve gives variation of temperature, and the lower curve the movement of the tree.

I have already shown : (1) that the diurnal movement just described is due to physiological reaction, and that the movement is abolished at the death of the plant; (2) that light has little or no effect, since the thick bark and bases of leaves screen the living tissue from the action of light ; (3) that transpiration has practically no effect on the periodic movement, since such movement takes place in other plants completely immersed under water; thus *Ipomœa aquatica*, a water plant, kept under water, gave the normal diurnal curve similar to that of the palm. The modifying effect of transpiration was in this case, completely excluded. I obtained similar effect with geotropically curved stem of *Basella cordifolia* (p. 25) ;

(4) that the weight of the plant-organ as such, has little effect on the diurnal curve, since an inverted plant continues for a few days to exhibit the periodic movement, in spite of the antagonistic effect of weight. A different experiment will be described (see p. 582) where the effect of weight was completely neutralised and the plant-organ gave, nevertheless, the normal diurnal curve.

I have also shown that the diurnal movement is determined by the modifying influence of temperature on

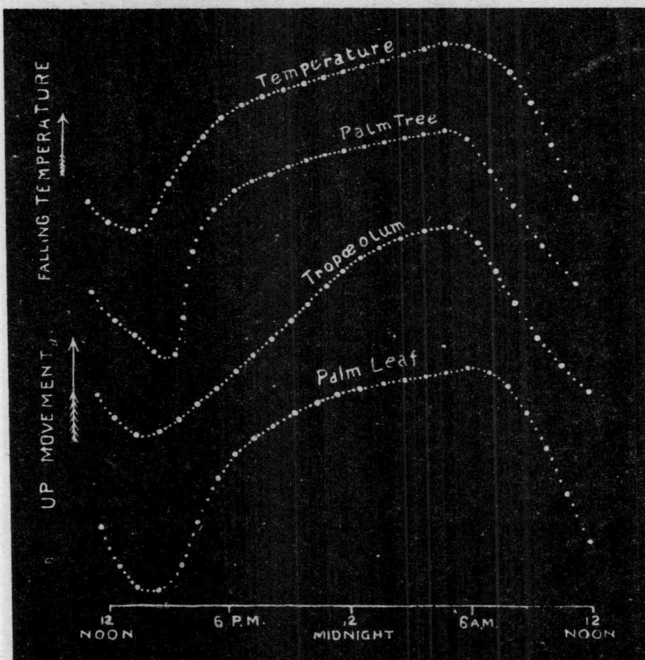

FIG. 200.—Diurnal record of inclined palm tree, of geotropically curved procumbent stem of *Tropæolum* and the dia-geotropic leaf of palm. Note general similarity between diurnal curve of plants and the thermographic record.

geotropic curvature. Rise of temperature opposes or neutralises the geotropic curvature; fall of temperature, on

the other hand, accentuates it. The particular diurnal movement was not confined to the palm trees, but was exhibited by all plant-organs subjected to the stimulus of gravity.

DIURNAL MOVEMENT OF PROCUMBENT STEMS AND OF LEAVES.

Experiment 210.—In order to demonstrate the continuity of the phenomenon of diurnal movement I took various stems

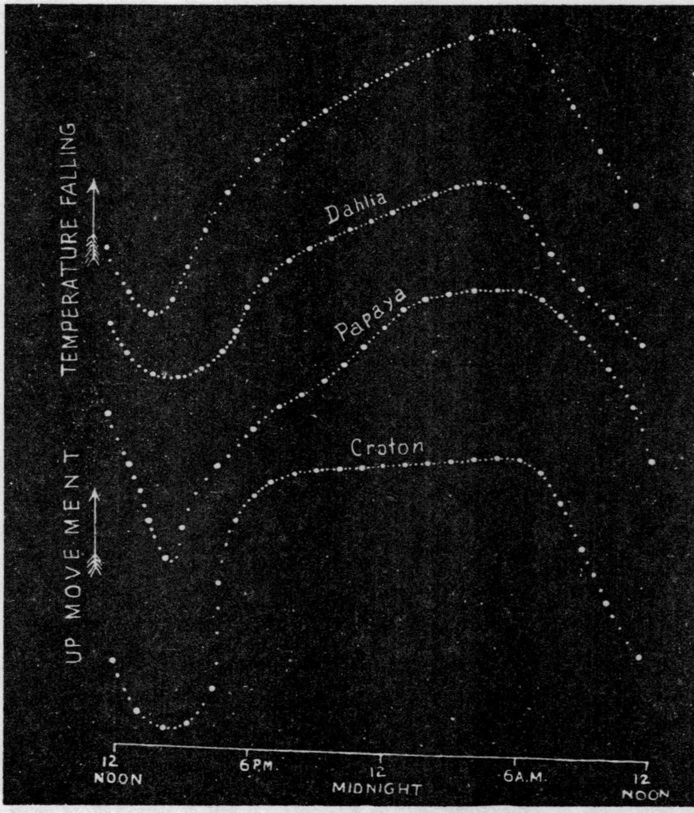

FIG. 201.—Diurnal records of leaves of *Dahlia, Papaya* and *Croton.*

growing in water or land for my experiment. The plants were laid horizontally, till the stems bent up and assumed the stable position of geotropic equilibrium. In figure 200 is given records of the inclined palm tree, of procumbent stem of *Tropæolum*, and the leaf of the palm tree. The very close relation between the temperature-variation and the movement of different plant-organs is sufficiently obvious.

I shall next give a series of diurnal records of leaves of different plants such as those of *Dahlia*, *Papaya* and *Croton* (Fig. 201). In these also fall of temperature induces an up-movement while rise of temperature causes a fall of the leaf. I shall presently refer to the 'personal equation' by which the record of one plant is distinguished from another.

CONTINUOUS DIURNAL RECORD FOR SUCCESSIVE THERMAL NOONS.

Experiment 211.—The diurnal record given above, was taken from ordinary noon at 12 o'clock to noon next day. The diurnal curve becomes much simplified if the record be taken from *thermal-noon* (at about 2 p.m.) to the thermal noon next day. The plant-organ becomes erected during falling temperature from thermal-noon to thermal-dawn next morning, and undergoes a fall during rise of temperature from thermal-dawn to thermal-noon. The subsequent diurnal records will therefore be given for 24 hours commencing with 2 p.m. In figure 202 is given diurnal records of geotropically curved stem of *Tropæolum* and the leaf of *Dahlia* for two days in succession.

The thermal record shows that there was a continuous fall of temperature from thermal-noon at 2 p.m. to the thermal-dawn at 6 a.m. next morning, that is to say, for 16 hours. Rise of temperature through the same range occurred in 8 hours from 6 a.m. till 2 p.m. The averages rate of rise of temperature was thus twice as quick as that of fall. This is clearly seen from the slopes of thermal curve during thermal ascent and descent. The recod of the movement of the plant shows a striking parallelism; the different plant-organs became erected from thermal-noon to thermal-dawn, and underwent a fall from thermal-dawn to thermal-noon. The descent of the curve is, as in the

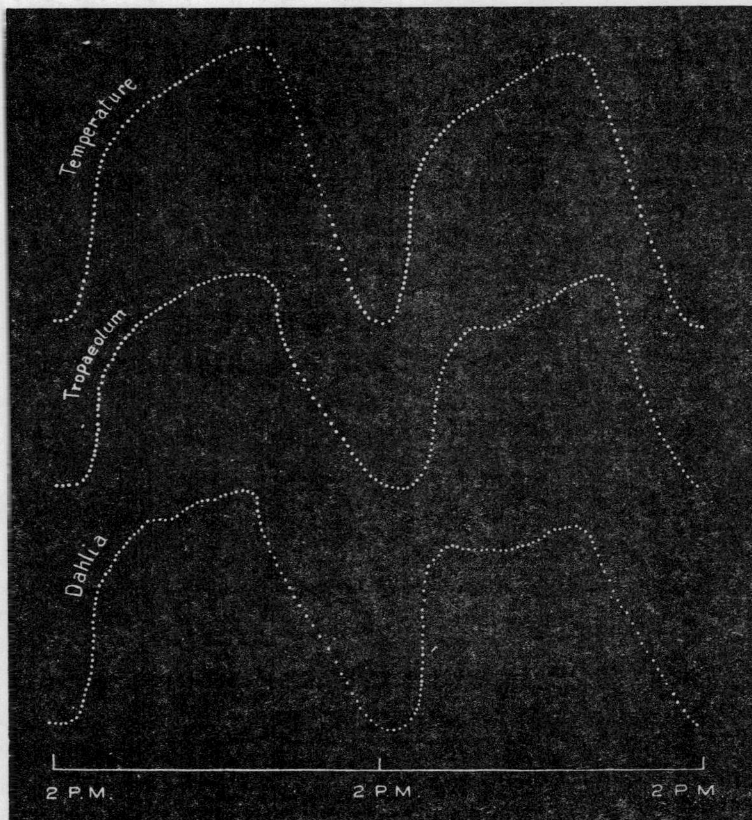

FIG. 202.—Diurnal curve of the procumbent stem of *Tropæolum majus*, and the leaf of *Dahlia* for two successive days. In the thermographic record the up-curve represents fall, and down-curve rise of temperature.

case of thermal curve, relatively more abrupt. The records on two successive days are very similar, the slight difference being due to the physiological depression consequent on prolonged maintenance of the plants in a closed chamber.

MODIFICATION OF THE DIURNAL CURVE.

I shall now proceed to explain the modifications that may occur in the standard thermo-geotropic curve.

Turning points.—In the bulky Palm, the reversal of movement from fall to rise or *vice versâ* takes place about an hour after the thermal inversion. This lag is partly due to the time taken by a mass of tissue to assume the temperature of the surrounding air. There is, moreover, the question of physiological inertia which delays the reaction. In leaves this lag may be considerably less or even absent. In certain cases the reversal of movement may take place a little earlier than the temperature inversion. It should be remembered in this connection, that in response to temperature change, the leaf is often displaced to a considerable extent from its 'mean position of equilibrium'; moreover the force of recovery is greatest at the two extreme positions. These considerations probably explain the quick return of the leaf to equilibrium position. The slow autonomous movement of the leaf may sometimes prove to be a contributory factor.

Effect of irregular fluctuation of temperature.—In settled weather the diurnal rise and fall of temperature is very regular. But under less settled condition, owing to the change of direction of the wind, the temperature curve shows one or more fluctuations, specially in the forenoon. It was a matter of surprise to me to find the plant-record repeating the fluctuations of thermal record with astonishing fidelity. This common twitch in the two records is seen in the record of the Sijberia Palm (Fig. 199). Certain plants are extremely sensitive to variation of temperature; so much so that these physiological indicators of thermal variation are far more delicate than ordinary thermometers.

Effect of restricted pliability of the organ.—A leaf is more pliable in one direction than in the other. The pulvinus of *Mimosa*, for example, allows a greater amount of bending downwards than upwards; in consequence of this the leaf in its fall becomes almost parallel to the internode below; the up-movement is, however, far more restricted. The leaf in its most erect position still makes a considerable angle with the internode of the stem above it. If the leaf-stalk of a plant be restricted in its rise the erectile movement at night will reach a limit, and the top of the curve will remain flat. This is seen illustrated in the record of the leaf of *Croton* (Fig. 202), which attains its maximum erection at 9 p.m. and the subsequent curve remains flattened till 7 a.m.; after this the leaf begins to execute its downward movement. In other cases, the range of up-movement is very great and the plant-organ erects itself continuously till morning. In certain cases the impulse of up-movement carries the organ beyond the stable position of equilibrium; after this the leaf begins to retrace its path slowly; the down-movement due to rise of temperature is, however, far more abrupt, and easily distinguishable from the previous slow return.

It will thus be seen that though the diurnal record consists of an alternating up and down curve, yet these minor characteristics or 'personal equation' of the plant confers on the record a certain stamp of individuality.

Effect of age.—In the floral leaves of *Nymphæa* the thermonastic movement is of positive sign; that is to say, an erection of the petal during rise, and a fall during the lowering of temperature. The corresponding movement of leaves would therefore be an erection of the leaf in day-time, and a fall of the leaf at night. The periodic curve of such leaves would be of opposite sign to the standard thermo-geotropic curves given above. The leaf of *Nicotina* is adduced as an example of a leaf which exhibits a movement of fall at night. But the fully grown and horizontally spread leaf I find that gives the normal record. The very young growing leaves give a different and somewhat erratic curve. The difference between growing and fully grown leaves is explained by the fact that the former would be affected by thermotropism, and the latter by thermo-geotropism. Young leaves exhibit moreover a pronounced hyponasfy or epinasty, which would naturally modify the diurnal curve.

Certain interesting variation is met with in the diurnal record of sprouting leaves of *Mimosa* in spring. The movements of leaves grown later in the season, as will be explained in a later chapter, are very definite and characteristic. But the young leaves in spring exhibit no definite diurnal curve, but a series of automatic pulsations, the unsuspected presence of which in all leaves of *Mimosa* will be demonstrated in a subsequent chapter. Later in the season, the leaf becomes tuned, as it were, to the periodic variation of the environment; the automatic movements become suppressed, and the diurnal periodicity becomes deeply impressed on the organism.

Effect of season.—The diurnal curve may also be modified by the seasonal variation of any one of the effective factors. *Tropæolum majus*, for example, exhibits positive phototropic action in one season and a negative reaction in a different season. These seasonal variations must necessarily modify the diurnal curve.

I shall now proceed to demonstrate the determining influence of thermal variation, and of stimulus of gravity on the thermo-geotropic movements. The striking similarity of the thermograph, and the record of movement of plants demonstrate the causal relation between temperature variation and diurnal movement, of which the two additional tests described below offer further confirmation.

REVERSAL OF NORMAL RHYTHM.

The normal diurnal movement is, as we have seen, a fall during rise of temperature from morning to afternoon, and a rise from afternoon till next morning. I succeeded in reversing the normal rhythm of *Basella* by reversing the normal variation of temperature at the two turning points, in the morning and in the afternoon. The plant was subjected to falling temperature in the morning and to rising temperature in the afternoon. The normal movement now became reversed, *i.e.*, an erection instead of fall in the forenoon and a fall instead of rise in the afternoon (p. 28).

EFFECT OF CONSTANT TEMPERATURE.

The second test which I shall employ is the effect of maintenance of constant temperature, which should wipe off, as it were, traces of periodic movement. It was necessary for this investigation to maintain the plant chamber at constant temperature throughout day and night. The usual thermostat is virtually a recess in a double-walled chamber filled with water, the chamber being covered with a heat insulating material. But this contrivance is unsuitable for the plant chamber which is to contain good sized plants, and the recording apparatus. The problem of maintaining a large air-chamber at constant temperature presented many difficulties which were ultimately overcome by the device of an extremely sensitive thermal regulator.

The Thermal Regulator.—I shall in a future paper give a complete account of the large thermostatic air-chamber. The important part of the apparatus is an electro-thermic regulator which interrupts the heating electric current as soon as the temperature of the chamber is raised a hundredth part of a degree above the predetermined temperature. The automatic make and break of the current takes place in rapid succession, and the temperature of the chamber is thus maintained constant within tenth of a degree, throughout day and night.

Diurnal record of Tropæolum *under constant temperature: Experiment 212.*—The normal record of geotropically curved *Tropæolum* is already given in figure 202. In repeating the record I maintained the plant at constant temperature for 24 hours; the result of this is seen in the first part of the record (Fig. 203). The thermal record

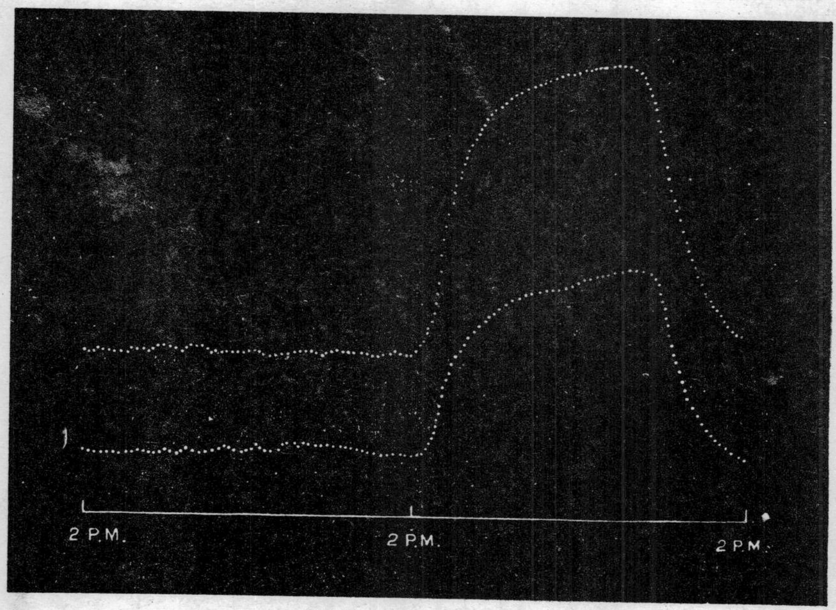

FIG. 203.—Abolition of diurnal movement in *Tropæolum* under constant temperature, and its restoration under normal daily fluctuation. The upper record is of temperature and the lower of plant movement.

is practically horizontal, and the diurnal record of the plant shows no periodic movement. The thermal regulator was on the next day put out of operation, thus restoring the normal diurnal variation of temperature. The record of the plant is seen to exhibit once more its normal periodic movement.

I have in the chapter on thermo-geotropism (p. 515) shown that the diurnal movement of a geotropically curved organ is determined in reference to the direction of force of gravity. This will be seen demonstrated in an interesting manner in the two following experiments on the effect of inversion of the plant on daily movement.

DIURNAL MOVEMENT IN INVERTED POSITION.

I have already referred to the distinction that is made between nastic and paratonic movements. In the former the movement is autonomous and in relation to the plant, and in the latter it is due to an external force which determines the direction of movement. In nastic reaction, closure movement would persist as a closure movement* ; but should the direction of movement be determined by the stimulus of gravity, closure movement would, on inversion, be reversed into an opening movement. Viewed from an external point of view an up-movement in the latter case would, after readjustment on inversion, become an up-movement, though in so doing, the expansion should be transferred from the upper to the lower side of the organ. It is to be understood in this connection, that some time must lapse before this readjustment is possible, and that the former movement may continue, in certain cases, as a persistence of after-effect.

I succeeded in demonstrating the paratonic effect of geotropic stimulus on the periodic movement of the palm leaf, by holding the plant in an inverted position (p. 24). On the first day of inversion, the diurnal record was erratic, but in the course of 24 hours, the leaf readjusted itself to its unaccustomed position, and became somewhat erected under geotropic action. After the attainment of

* By closure is meant movement of opposite pairs of leaf-organs towards each other.

this new state of geotropic equilibrium, the leaf gave the record of down-movement during rise, and up-movement during fall of temperature, movements which in reference to the plant are the very opposite to those in a normal position. But seen from an external point of view, rise of temperature caused in both normal and inverted positions, a down-movement indicative of diminished geotropic curvature; fall of temperature, on the other hand, brought about an erectile movement, thus exhibiting enhancement of geotropic curvature.

Experiment 213.—A still more striking result exhibiting the phase of transition was given by the geotropically

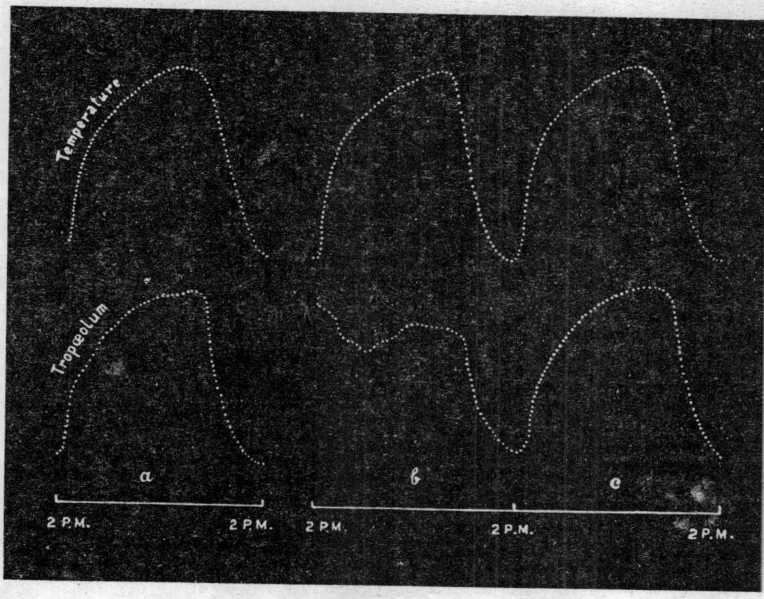

FIG. 204.—Effect of inversion of the plant on diurnal movement. (*a*) Normal record, (*b*) record 24 hours after inversion and (*c*) after 48 hours (*Tropæolum*).

curved stem of *Tropæolum*. Its diurnal curve and the subsequent changes after inversion are given in figure 204. In (*a*) is seen the normal diurnal curve; the specimen was inverted, and it took an entire day for the plant to readjust itself to the new geotropic condition. The record (*b*) was recommenced on the second day after inversion; the persistence of previous movement is seen in the reversed curve during the first half of the second day; but in the second half the record became true, and the third day the inverted plant gave a record which, from an external point of view, was similar to that given by the plant in the normal position.

SUMMARY.

A continuity is shown to exist between the thermo-geotropic response of rigid trees, stems, and leaves of plants.

The diurnal record exhibits an erectile movement from thermal-noon to thermal-dawn, and a movement of fall from thermal-dawn to thermal-noon.

In contrast with thermonastic movement which takes place in growing organs, thermo-geotropic movement takes place in fully grown organs including rigid trees. The thermonastic movement is independent of the direction of gravity, while in thermo-geotropic reaction, the stimulus of gravity exerts a directive action.

The effect of variation of temperature on the diurnal movement is demonstrated by induced change of normal rhythm, by artificial transposition of periods of thermal inversion, and by the abolition of periodic movement under constant temperature.

The effect of stimulus of gravity on the diurnal movement is demonstrated by the effect induced on holding the plant upside down. The direction of the daily movement is found to be determined by the directive action of the stimulus of gravity.

L.—THE AFTER-EFFECT OF LIGHT

By

SIR J. C. BOSE,

Assisted by

SURENDRA CHANDRA DAS.

WE have considered two types of diurnal movement, one due to the predominant action of variation of light, and the other, to that of changing temperature. There are, however, other organs which are sensitive to variations both of light and of temperature. The effect of light is, generally speaking, antagonistic to that of rise of temperature; hence the resultant of the two becomes highly complex.

Still greater complexity is introduced by the different factors of immediate and after-effect of light. This latter phenomenon is very obscure, and I attempted to determine its characteristics by electrical method of investigation. A fuller account of after-effect of light on the response of various plant-organs and of animal retinæ will be found elsewhere.* I shall here refer only to one or two characteristic results which have immediate bearing on the present subject.

Direct stimulation under light induces excitatory reaction, which is mechanically exhibited by contraction, and electrically by induced galvanometric negativity. Under continuous stimulation, the excitatory effect, either of positive curvature or of induced galvanometric negativity, is found to attain a maximum. This is often found to undergo a decline and reversal; for under continuous

* "Comparative Electro-Physiology"—p. 392.

stimulation there is a fatigue-decline, as seen in the relaxation following normal contraction in animal muscle. The positive tropic curvature, and the induced galvanometric negativity may thus undergo a decline, and neutralisation. This neutralisation is also favoured, in certain cases, by transverse conduction of excitation to the distal side.

The character of the after-effect will presently be shown to be modified by the duration of previous stimulation, the different phases of which will for convenience, be distinguished as pre-maximum, maximum and post-maximum. Since stimulus simultaneously induces positive "A" and the negative "D" changes (p. 143), their intensities will undergo relative variation during the continuance and cessation of stimulus. The after-effect will therefore exhibit unequal persistence of the expansive "A" and contractile "D" reaction at different phases of stimulation.

ELECTRIC AFTER-EFFECT.

Confining our attention to the electric response, it is found that under continued action of light the excitatory galvanometric negativity increases to a maximum, after which there is a decline, and neutralisation. Figure 205 gives the galvanographic record of the electric response of the leaf stalk of *Bryophyllum* under light; the up-curve represents increasing negativity which, after attaining a maximum, undergoes neutralisation as seen in the down-curve. I shall, with the help of the diagram given in the next figure, describe and explain the various after-effects I observed on sudden stoppage of light: before the attainment of maximum, at the maximum, and after the maximum.

After-effect of pre-maximum stimulation: Experiment 214.—Light is applied at arrow and stopped in different experiments at *a*, *b*, and *c* (Fig. 106). Continuous stimulation induces increasing galvanometric negativity ; when

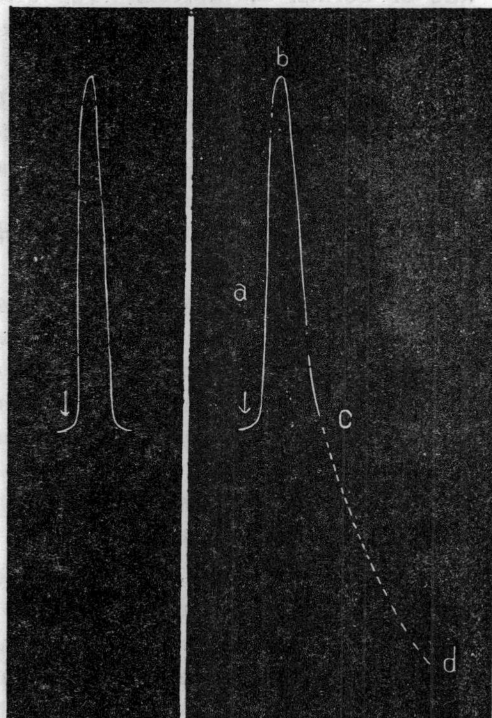

FIG. 205. FIG. 206.

FIG. 205.—Electric response of the leaf-stalk of *Bryophyllum* under continuous photic stimulation. Increasing negativity represented by up-curve, neutralisation by down-curve.

FIG. 206.—Diagrammatic representation of electric after-effect of stimulation. Pre-maximal stimulation produced by stoppage of light at *a*, gives rise to continuation of previous response followed by recovery. Stoppage of light at maximum *b* gives rise to recovery to equilibrium position. Stoppage of light at post-maximum *c*, gives rise to over-shooting below zero line.

stimulus is stopped at *a* before the maximum, the after-effect is a persistence of excitatory galvanometric negativity, which carries the response record higher up; after a certain interval recovery takes place and the record returns to the zero line of normal equilibrium. The after-effect of pre-maximum stimulation is thus a short-lived continuance of response followed by recovery.

After-effect at maximum: Experiment 215.—In this the photic stimulus was continued till the attainment of maximum, when light was suddenly removed at *b*. The after-effect was no longer a persistence of responsive movement, but disappearance of negativity and recovery to zero line of equilibrium.

Post-maximum after-effect: Experiment 216.—In this light was continued till there was a complete neutralisation, the curve of response returning to zero line; to all outer seeming the responsive indication of the tissue is the same as before excitation. But stoppage of stimulus at *c* causes an over-shooting at a rapid rate far *below* the zero line; and it is after a considerable period that the curve returns to the zero line of equilibrium.

The condition at post-maximum *c* is thus one of dynamic equilibrium where two opposite activities, "A" and "D," balance each other; for had the condition of the 'neutralised' tissue been exactly the same when fresh, cessation of stimulus would have kept the galvanometric spot of light at the zero position.

The electric investigation described above shows that the after-effect is modified by duration of stimulation, and that:

(1) the after-effect of pre-maximum stimulation is the continuation of response in the original direction (upward, and away from zero line), followed by recovery,

(2) the after-effect of the maximum is an electric recovery towards zero position, and

(3) the after-effect of post-maximum stimulation is an over-shooting *downward* below the zero line.

TROPIC RESPONSE UNDER LIGHT AND ITS AFTER-EFFECT.

I shall now describe the after-effect of light as seen in mechanical response, and the results will be found parallel to those given by the electric response. The specimen employed is the terminal leaflet of *Desmodium gyrans*, the pulvinus of which is very sensitive to light. Pulvinated organs, generally speaking, exhibit a diurnal variation of turgor in consequence of which the position of equilibrium of the leaf or leaflet undergoes a periodic change. But this equilibrium position of the organ remains fairly constant for nearly two hours about mid-day, the variation of temperature at this period being slight. We may therefore obtain the pure effect of light by carrying out the experiment at this period, and completing it within a short time to avoid complication arising from the autonomous variation of turgor.

The period of experiment of the plant may be shortened by a choice of suitable intensity of light; a given tropic effect induced by prolonged feeble light may thus be obtained by short exposure to stronger light. The source of light for the following experiment was a 50 c.p. incandescent lamp. The intensity was increased to a suitable value by focussing light on the upper half of the pulvinus by means of a lens. The intensity was so adjusted that the maximum positive curvature was attained in the course of about 6 minutes, and complete neutralisation after an exposure of 17 minutes.

Pre-maximum after-effect: Experiment 217.—Light was allowed to act on the upper half of the pulvinus for two minutes and twenty seconds; this induced an up-movement *i.e.*, a positive curvature. On the stoppage of light the up-movement continued for one minute and twenty seconds, after which the down-movement of recovery was completed in six minutes and twenty seconds (Fig. 207). The immediate after-effect is thus a movement upward, away from the

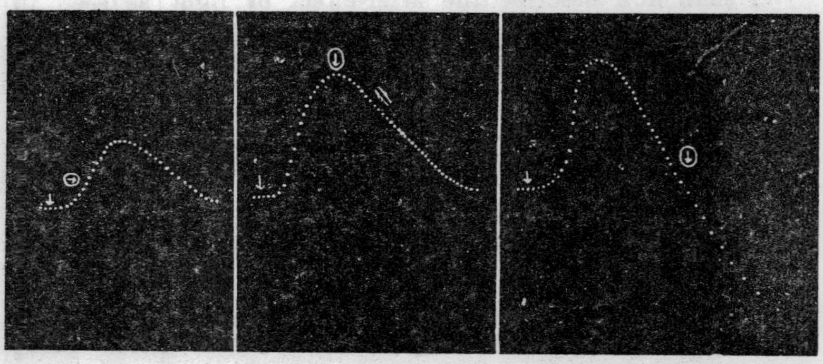

FIG. 207. FIG. 208 FIG. 209.

FIG. 207.—Light applied at arrow, and stopped at the second arrow within a circle. After-effect of pre-maximum stimulation is continuation of positive curvature followed by recovery.

FIG. 208.—After-effect at maximum; recovery towards zero position of equilibrium.

FIG. 209.—After-effect at post-maximum is a rapid overshooting below the position of equilibrium. Light was applied in all cases on upper half of pulvinus of terminal leaflet of *Desmodium gyrans*.

zero line of equilibrium. The result is seen to be the same as the electric after-effect of pre-maximum stimulation.

After-effect at maximum: Experiment 218.—Application of light for 5 minutes and twenty seconds induced a maximum positive curvature. Stoppage of light was followed at once by recovery which was completed in about 10 minutes (Fig. 208).

After-effect at post-maximum: Experiment 219.—As the plant was fatigued by previous experiments, a fresh specimen was taken and light was applied continuously on the upper half of the pulvinus. This gave rise first to a maximum positive curvature, subsequently diminished by transverse transmission of excitation. Neutralisation took place after application of light for 17 minutes. On the stoppage of light, there was a sudden overshooting *below* the zero line (Fig. 209), and the rate of the movement on the cessation of light was nearly twice as quick as during the process of neutralisation.

SUMMARY.

The after-effect of light is modified by the duration of exposure to light.

Under continued action of light, the electric response of galvanometric negativity in plants attains a maximum after which it undergoes decline, and neutralisation.

The electrical after-effect exhibits characteristic differences depending on the duration of previous exposure to light.

The pre-maximal after-effect is a temporary continuation of response under light followed by recovery.

The after-effect at the maximum is a recovery to the normal equilibrium.

The after-effect at post-maximum is an 'overshooting' below the position of equilibrium.

The immediate and after-tropic response of light are similar to the corresponding photo-electric effects.

The pre-maximum after-effect is a continuation of positive tropic movement followed by recovery; the after-effect at maximum is a recovery to the normal equilibrium position of the organ. The post-maximum after-effect is an overshooting below the position of normal equilibrium.

LI.—THE DIURNAL MOVEMENT OF THE LEAF OF *MIMOSA*

By

SIR J. C. BOSE.

IN the standard curve of nyctitropic movement under thermo-geotropism described in a previous paper, the diurnal record consisted of an up-curve from thermal-noon to thermal-dawn, and a down-curve from the thermal-dawn to thermal-noon. The responding organ, which may be an inclined stem or a horizontally spread petiole, underwent an erection during the decline of temperature, and a fall with the rise of temperature. The diurnal record of the *Mimosa* leaf appears, however, to be totally different.

Experiment 220.—I obtained the diurnal record of *Mimosa* (Fig. 210) for twenty-four hours commencing at 2 p.m. which is the thermal-noon. The summer and winter records are essentially the same ; the only difference is in the greater vigour of movement exhibited by summer specimens. The diurnal movement of the leaf is very definite and characteristic ; for the curves taken five years

ago do not differ in any way from those obtained this year. The record may conveniently be divided into four phases.

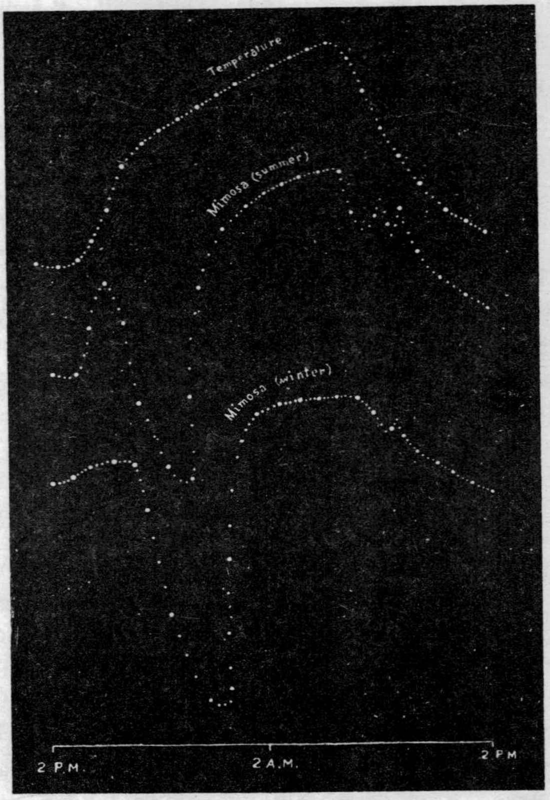

FIG. 210.—Diurnal record of *Mimosa* in summer, and in winter. Leaf rises from 2 to 5 p.m., when there is a spasmodic fall. Leaf re-erects itself from 9 p.m. to 6 a.m. after which there is a gradual fall till 2 p.m. with pulsations. The upper-most record gives temperature variation, up-curve representing fall of temperature and *vice versâ*.

First phase.—The leaf erects itself after the thermal-noon up to 5 or 5-30 p.m. The temperature, it should be remembered, is undergoing a fall during this period.

Second phase.—There is a sudden fall of the leaf in the evening which continues till 9 p.m. or thereabout.

Third phase.—The leaf erects itself till thermal-dawn at about 6 a.m. next morning.

Fourth phase.—There is a fall of the leaf during the rise of temperature from thermal-dawn to thermal-noon. The uniformity of the fall is, however, interrupted by one or more pulsations in the forenoon. These pulsations are more frequent in summer than in winter.

It will thus be seen that the difference between the normal thermo-geotropic curve, and the curve of *Mimosa* is not so great as appears at first sight. With the exception of the spasmodic fall in the evening, the diurnal curve shows an erectile movement during lowering of temperature, and a movement of fall during rise of temperature. I shall presently explain the reason of the sudden fall in the evening, and of the multiple pulsations in the forenoon.

I have, moreover, been able to trace a continuity in *Mimosa* itself, between the standard thermo-geotropic reactions and the modification of it by the action of light. The young leaves which sprout out at the beginning of spring take some time to become adjusted to the diurnal variation. There are two intermediate stages through which the leaves pass before they exhibit their characteristic diurnal curve. Slow rhythmic pulsations are at first seen to occur during day and night. At the next stage the leaves exhibit the diurnal movement of fall from thermal-dawn to thermal-noon, and movement of erection from thermal-noon to thermal-dawn next morning, the record being in every way similar to the standard thermo-geotropic curve. It is only at the final stage that there is a spasmodic fall in the evening which we shall find is the characteristic after-effect of light.

Before proceeding further I shall refer briefly to the theory of Millardet in explanation of the diurnal movement of the leaf of *Mimosa*. He found that the tension in stems, and presumably its turgor, is increased with rise and decreased with fall of temperature. The movement of the lateral leaf may, therefore, be due to the induced variation of tension in the main axis. Had this been the case the minimum tension would have occurred at the minimum temperature in the morning, and the leaf should have undergone a maximum fall. The maximum temperature attained in the afternoon should have, on the other hand, brought about the maximum erection. The observed facts are, however, the very opposite to these. Kraus and Millardet also found that light and darkness had great influence on the tension, which increases in darkness and diminishes in light. The tension at dawn may therefore be a resultant of the depressing effect of low temperature opposed by the promoting effect of darkness, the latter being the predominant factor. The erect position of *Mimosa* leaf in the morning may thus be accounted for by the resultant increase of tension of the stem. The explanation of the movements of the leaves is thus to be attributed to the variation of tension in the main axis to which the leaves are attached; this leads to the conclusion that the leaf-movement should be determined in relation to the plant, and not in relation to the external stimulus. I shall, however, describe a crucial experiment in the course of this paper, which will show that the direction of stimulus of gravity has a determining influence on the periodic movement. The sudden fall of the leaf before evening is again inexplicable from the theory of periodic variation of tension.

The complexity in the diurnal movement in *Mimosa* arises from the fact that there are three factors whose fluctuating effects are different at different parts of the day. The effect at any particular hour results from the algebraical summation of the following factors: (1) the thermo-geotropic action, (2) the immediate effect of photic stimulus and (3) the after-effect of light. The leaf of *Mimosa* has, moreover, as I shall show, an autonomous movement of its own. I shall take up the full consideration of the subject in the following order:

1. *The thermo-geotropic reaction.*—A crucial experiment will be described which demonstrates the effect of thermo-geotropism in the diurnal movement of the leaf of *Mimosa.*

2. *Autonomous pulsation of* Mimosa.—The natural pulsation of the plant is obscured by the paratonic effect of external stimuli. I shall explain the method by which the natural pulsation of the leaf becomes fully revealed.

3. *The immediate effect of light.*—This is not constant, but will be shown to undergo a definite variation with the intensity and duration of light. A very great difficulty in the study of effect of daylight at different parts of the day is introduced on account of the absence of any reliable recorder for measurement of fluctuation of light. I shall describe a device which gives a continuous record of photic variation for the whole day.

4. *The after-effect of light.*—The spasmodic fall of the leaf of *Mimosa* towards the evening presents the most difficult problem for solution. I shall first describe the diurnal movement of another plant which presents characteristics similar to those of *Mimosa.* I shall also demonstrate the various after-effects of light at different parts of the day. These results will offer the fullest explanation of the sudden fall of the leaf towards evening.

As regards the sudden fall of the leaf about evening, Pfeffer regarded it as due to increased mechanical moment of the secondary petioles moving forward on the withdrawal of light. I shall, however, in the course of this paper show, that the characteristic movements occur even after complete removal of the sub-petioles. In the following experiment, carried out with the intact plant, the effect of possible variation of weight is completely eliminated. In spite of this, the diurnal movement exhibited its characteristic phases including sudden movement in the evening.

The experiment I am going to describe will exhibit the diurnal curve obtained by an entirely different method, and will clearly exhibit the thermo-geotropic effect, as well as the immediate and after-effect of light.

DIURNAL VARIATION OF GEOTROPIC TORSION.

I have shown that the pulvinus of *Mimosa*, subjected laterally to the action of stimulus of gravity, exhibits a torsional response. When the *Mimosa* plant is laid sideways, so that the plane of separation of the upper and lower halves of the pulvinus is vertical, geotropic stimulus acts laterally on the two halves of the differentially excitable pulvinus. When the less excitable upper half is to the left of the observer (see Fig. 179), the responsive torsion under geotropic stimulus will be clock-wise, the less excitable upper half of the pulvinus being thereby made to face the vertical lines of gravity. When the plant is turned over to the other side (the less excitable upper half being now to the right of the observer) the induced torsion will be counter clock-wise. The response is therefore determined by the directive action of stimulus of gravity. Light has also been shown to give rise to torsion (p. 400). Light acting in the same direction as the stimulus of gravity, *i.e.*, from above, enhances the rate of torsion, the curve of response being due to the joint effects of light and gravity.

Experiment 221.—I obtained 24 hours' record of variation of torsional response of *Mimosa*, commencing with thermal-noon at 2 p.m. It is to be borne in mind that increase of torsion indicates increase of geotropic action, just as the erectile movement of the leaf in the normal position indicates the enhanced geotropic effect. Inspection of figure 211 shows that the fall of temperature after thermal-noon was attended by increase of torsion. The

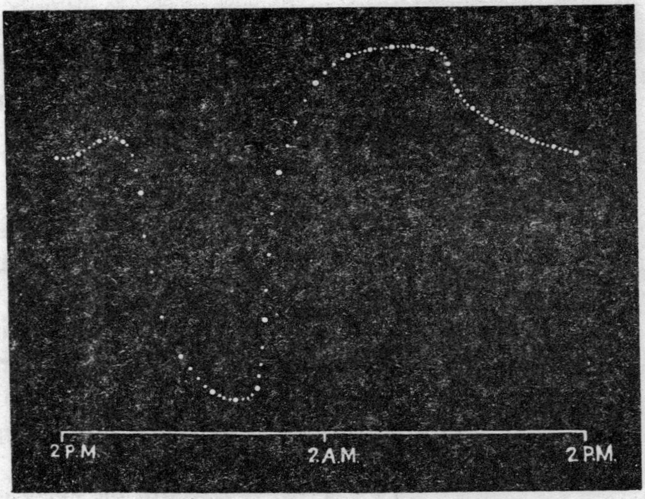

FIG. 211.—Record of diurnal variation of torsion in *Mimosa* leaf. Up-curve represents increase and down-curve decrease of geotropic torsion.

curve went up till about 5 p.m., as in the ordinary record of *Mimosa*. The torsion suddenly decreased with the rapid diminution of light after 5 p.m. The torsion then increased with falling temperature from 9 p.m. till thermal-dawn next morning. After 6 a.m. there is a continuous diminution of torsion till 5 p.m.

We may now summarise the diurnal variation of torsion exhibited by *Mimosa*. The torsion undergoes a periodic increase during the fall of temperature from afternoon till next morning, and a diminution during rising temperature from morning till afternoon. A sudden diminution of torsion occurs at about 5 p.m. due to the disappearance of light. The torsional record is, to all intents and purposes, a replica of the record of periodic up and down movements of the leaf.

This method of torsion has several advantages over the ordinary method. First, the petiole being supported by the loop of wire, the weight of the leaf has no effect on the curve of response. In the second place, the periodic variation of turgor of the stem, as suggested by Millardet, will not in any way affect the record. Variation of turgor can only cause a swing to and fro, in a direction perpendicular to the plane which divides the pulvinus into upper and lower halves; it can in no way induce a torsional movement, or a variation of the rate of that movement.

The automatic pulsation of the leaf of Mimosa.—The occurrence of the pulsatory response in the morning record of *Mimosa* led me to search for multiple activity in the response of the pulvinus. I have in my previous investigation on the electric response of *Mimosa* obtained multiple series of responses to a single strong stimulus. Blackman and Paine have recently shown that an isolated pulvinus of *Mimosa* exhibit multiple mechanical twitches under excitation.*

Even under normal conditions, the sprouting young leaves in March, as already stated, exhibit automatic pulsations throughout the day and night; in older leaves tuned to diurnal periodic movements, these natural pulsations are more or less suppressed. But in the forenoon, several pulsations are exhibited even by the old leaves.

* Blackman and Paine—"Annals of Botany" January 1918.

The question may now be asked: Why should the pulsations occur preferably in the morning? In connection with this I shall refer to the suppression of the pulsatory activity of *Desmodium gyrans* when the leaflet was pulled up by the action of light (cf. Fig. 188). The leaf of *Mimosa* executes a very rapid movement of erection at night, and the natural pulsations are thereby rendered very inconspicuous. These pulsations may, however, be found in the night record of young leaves. The general occurrence of pulsations in the forenoon is probably due to the fact that the resultant force which causes the down-movement is at the time relatively feeble—the operative factors being: (1) the action of the rising temperature which induces down-movement, and (2) the action of light which in the forenoon opposes this movement. It will thus be seen that the forces in operation in the forenoon are more or less in a state of balance, hence conditions for exhibition of natural pulsations are more favourable in the morning than in other parts of the day.

Experiment 222.—I next tried to discover conditions under which the plant would exhibit its normal rhythmic activity during the whole course of 24 hours. The external stimuli which may interfere with the exhibition of its automatic pulsations are those due to gravity and light. They act most effectively on the pulvinus, when that organ is more or less horizontal and therefore at right angles to the direction of the incident stimulus; they act least effectively on the pulvinus when the organ is parallel to the direction of the external force. This latter condition may be secured by holding the plant upside down, when the pulvinus bends up and the leaf becomes erect and almost parallel to the vertical lines of gravity and to vertical light from above. The leaf, now relatively free from the effects of external stimulus, was found to exhibit its autonomous pulsations for more than seven days. I reproduce two sets of records (Fig. 212) for 24 hours each,

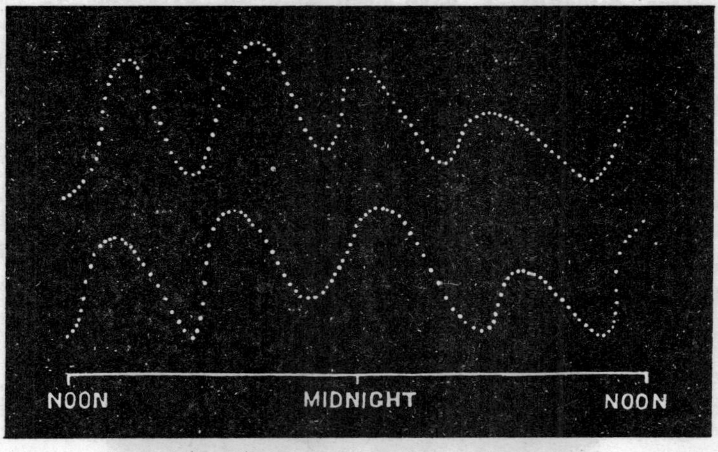

FIG. 212.—Continuous record of automatic pulsation of *Mimosa* leaf. The two series are for the first and the third day.

obtained on the first and the third day. The average period of a single pulsation is slightly less than six hours; but this is likely to be modified by the age of the specimen and the temperature of the environment.

One of the factors that determines the diurnal movement of the leaf is the immediate and after-effect of light. The movement under the action of light, is modified by the intensity and duration of illumination. The experimental investigation of the subject offers many difficulties, principally owing to the absence of any reliable indicator for the varying intensity of light during the course of the day.

THE PHOTOMETRIC RECORDER.

This difficulty I have been able to overcome by the automatic device for continuous record of the variation of light. The electric resistance of a selenium cell undergoes diminution with the intensity of light that falls on it. The photo-sensitive cell was made the fourth arm of a Wheatstone bridge, the resistance of the cell being exactly balanced when the shutter of the sensitive cell was closed. The selenium receiver was pointed upwards against the sky. Precaution was taken that it was protected from the direct action of sunlight. On opening the shutter a deflection of the index of a sensitive galvanometer was produced,

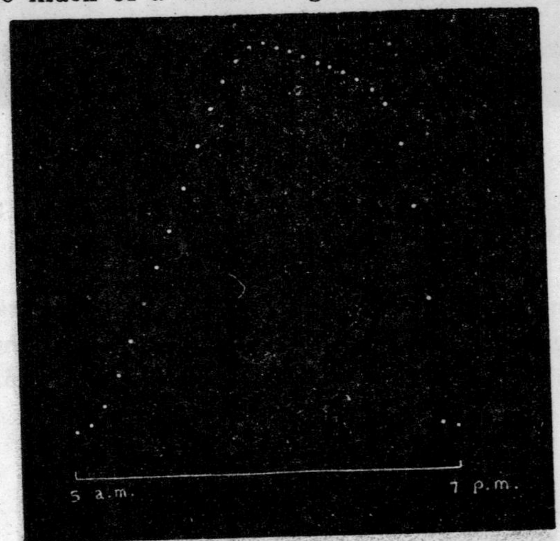

FIG. 213—Photometric record showing variation of intensity of light from morning to evening. Successive dots are at intervals of 30 minutes.

and the deflection increased with increasing intensity of diffuse skylight. The special difficulty was in securing automatic record of the galvanometer deflections. This was obtained by a special contrivance of an oscillating smoked glass plate, the up and down oscillation being at intervals of 30 minutes. A detailed account of this apparatus will, with its possibilities for meteorology, be given in a future paper. I reproduce the record obtained in my greenhouse on the 5th March (1919), which gives a general idea of the variation of the light from morning to evening (Fig. 213). The record shows that the light began to be perceptible at 5-30 a.m., and that the intensity increased rapidly and continuously till it reached a climax at noon, after which it began to decline slowly. The decline of intensity of light was very abrupt after 5 p.m., the effect being reduced to zero at 6-30 p.m.

THE EFFECT OF DIRECT LIGHT.

Under natural conditions, the leaf of *Mimosa* is acted on by light from above, and it is generally supposed that the pulvinus is positively phototropic, that is to say, it curves upwards till the leaf is placed at right angles to the direction of light. My investigations show, however, that the phototropic effects vary from positive to negative through an intermediate stage of neutralisation, these depending on the intensity and duration of exposure. When light acts continuously on the upper half of the pulvinus, there follows the following sequences of reaction :

(1) The leaf is at first erected by the contraction of the upper half of the pulvinus due to direct action of light acting from above.

(2) Under continuous stimulation of the upper half of the pulvinus by light, the excitation is slowly conducted to the lower half across the pulvinus. In consequence of this transmitted excitation, the lower half begins to contract and thus neutralises the first effect of erection. The upper half of the pulvinus is less contractile than the lower

half, and the neutralisation is due to the full contraction of the upper half antagonised by slight contraction of the lower half. The horizontal position of the leaf under light is therefore the result of balance of the two antagonistic reactions. If the incident light be very strong, the more intense transmitted excitation induces greater contraction of the lower half, and bring about a resultant down-movement (*cf.* p. 331).

Let us consider the effect of daily variation of light on *Mimosa*; we have here to take account both of intensity and duration. The intensity of light is seen to undergo a continuous increase which reaches a climax at noon; it then begins to decline slowly and the diminution of intensity of light is very abrupt after 5 p.m.

Under natural conditions the following phototropic effects are observed during the course of the day: light acting from above induces an up-movement of the leaf; but this is opposed by the thermo-geotropic fall of the leaf due to rise of temperature. As the two opposing effects are nearly balanced, any fluctuation of the relative intensity of the two gives rise to the pulsatory movements often seen in the forenoon; the *Mimosa* leaf has moreover an autonomous movement of its own. Under continued action of light neutralisation begins to take place after 1 p.m. (*cf.* *Expt.* 135). Later in the day the phototropic effect may become negative; reversal into this negative takes place under the joint action of intensity and duration of light; it takes place earlier under strong, and later under feeble, light.

THE EVENING SPASMODIC FALL OF THE LEAF.

I shall now deal with the difficult problem of the sudden fall of the leaf after 5 p.m. Pfeffer regarded this sudden fall in the evening as due to the increased mechanical moment of the secondary petioles moving forward on the withdrawal of light. But the following experiment shows that the increased mechanical moment cannot be the true explanation of the fall.

Diurnal movement of the amputated petiole: Experiment 223.—In my present experiment the possibility of variation of mechanical movement was obviated by cutting off the end of the petiole, which carried the sub-petioles. The cut end was coated with collodion flexile to prevent evaporation. The intense stimulus caused by amputation induced the excitatory fall of the leaf, but it recovered its normal activity after a period of three hours or so. The diurnal record of the leaf was commenced shortly after 1 p.m.; it will be noticed that the leaf, though deprived of the weight of its sub-petioles, still exhibited a sudden fall at about 5 p. m. (Fig. 214). The fall of the leaf cannot therefore be due to increased mechanical moment. The effect of weight was, moreover, eliminated in torsional response (*Expt.* 221). In spite of this the leaf exhibited a sudden movement after 5 p.m.

FIG. 214.—Record of leaf of *Mimosa* after amputation of sub-petioles. The leaf fell up to 2-30 p.m., and rose till 5 p.m., after which there is a spasmodic fall. (Successive dots at intervals of 15 minutes.)

Pfeffer has in his 'Entstehung der Schlafbewegung' (1907) offered another explanation of the sudden fall of the leaf of *Mimosa*. This, according to him, is not the direct effect of diminished intensity of light in the evening, but is due to the release of the leaf from the phototropic action of light, which, so long as it is sufficiently intense, holds the leaf in the normal position with its upper surface at right angles to the incident rays. Thus, on being set free from the strong action of light, the leaf moves in accordance with the preceding condition of tension; and as this is low the leaf falls, soon to rise again as the tension increases in prolonged darkness.

The above explanation presupposes: (1) that the tension was continuously decreasing till the evening, and (2) that as soon as the phototropic restraint which held the leaf up was removed it fell down in accordance with the prevailing diminished tension.

Referring to the first point, an inspection of the diurnal curve of *Mimosa* shows that the leaf had no natural tendency to fall towards the evening. There was on the contrary a movement of erection, on account of fall of temperature after the thermal-noon (Fig. 210). As the natural tendency of the leaf was to erect itself, the removal of phototropic restraint cannot therefore induce a movement of fall.

As regards the factor of light, the effect in the afternoon is a down-movement on account of transverse conduction of excitation; but the leaf is prevented from exhibiting this down-movement by the thermo-geotropic up-movement due to fall of temperature after the thermal noon. I shall presently describe experiments on the pure effect of light, which will show that the action of continued photic stimulus induces a down-movement of the leaf in the afternoon.

The results of experiments that have been described show that the sudden fall of the leaf in the evening could not be due to:

(1) increased mechanical moment,

(2) the natural tendency of the leaf to fall towards evening against phototropic action by which the leaf is held up.

The above explanations having proved unsatisfactory we have to search for other factors to account for the fall of the leaf on the cessation of light. In this connection I was struck by the extraordinary similarity of the diurnal curve of the petiole of *Cassia alata* with that of *Mimosa.*

DIURNAL CURVE OF THE PETIOLE OF *Cassia alata.*

Experiment 224.—The leaf of *Cassia* exhibits as in the leaf of *Mimosa* a slight erectile movement after the thermal-noon at 2 p.m., there is next a sudden fall after 5 p.m., which continues about 9 p.m.; after this the leaf exhibits a continuous rise with the fall of temperature, till the climax is reached about 6 a.m. in the morning; the leaf then

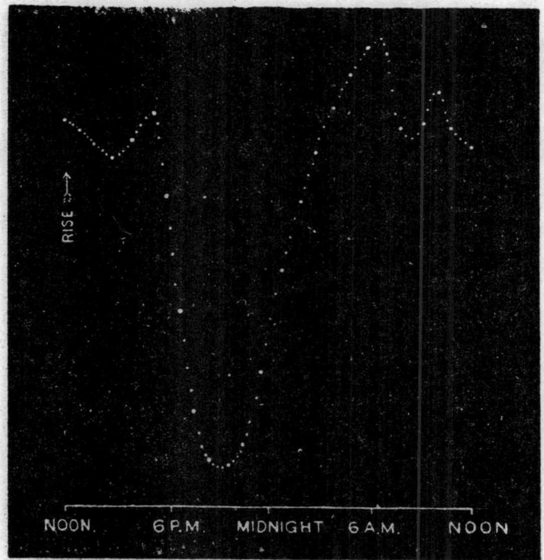

FIG. 215.—Diurnal record of *Cassia* leaf. Note similarity with diurnal record of *Mimosa*.

undergoes a fall with rise of temperature, there being a number of pulsatory movements in the forenoon, evidently due to unstable balance under the opposing effects of light and of rise of temperature (Fig. 215).

The reason of this similarity between the records of *Cassia* and *Mimosa* was found in the fact :

(1) That the main pulvinus of the leaf of *Cassia* is, like the pulvinus of *Mimosa*, differentially excitable, the lower half being more excitable than the upper. This is demonstrated by sending a diffuse electric shock through the leaf, the response being by a fall of the leaf due to the greater contraction of the lower half of the pulvinus. The leaf recovered after an interval of 20 minutes, the curve of response being similar to that of *Mimosa*. The only difference between the two organs is in the lesser excitability of the pulvinus of *Cassia*, on account of which a greater intensity of shock is necessary for producing the responsive fall.

(2) The responses to light are the same in both as will be seen in the following experiment.

Experiment 225.—In *Cassia*, as in *Mimosa*, light acting

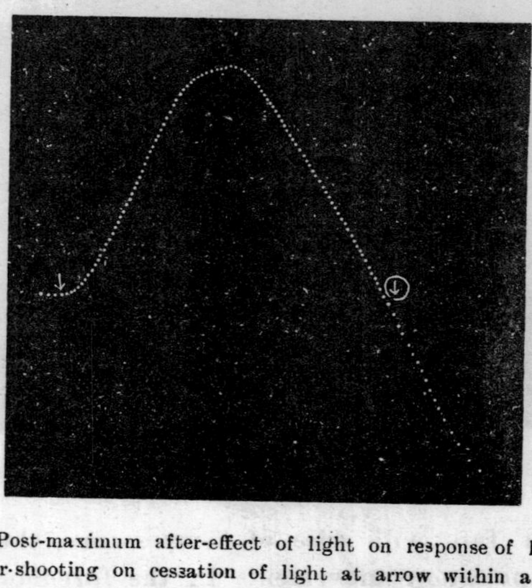

FIG. 216.—Post-maximum after-effect of light on response of leaf of *Cassia*. There is an over-shooting on cessation of light at arrow within a circle.

from above induces at first an erectile movement which reaches a maximum ; after this there is a neutralisation and reversal. In the record given in figure 216, light from a small arc lamp acting on the upper half of the pulvinus for 48 minutes gave the maximum positive curvature; this was completely neutralised by further exposure to light for 20 minutes. Light was cut off at neutralisation and there was a sudden fall beyond the equilibrium position, which was more rapid than the movement under light. The after-effect of prolonged exposure is thus an 'over-shooting' beyond the normal position of equilibrium.

RESPONSE OF *Mimosa* TO DARKNESS AT DIFFERENT PARTS OF THE DAY.

I now tried the effect of darkness on the movement of *Mimosa*, and was surprised to find that while artificial darkness caused a sudden fall of the leaf in the afternoon, it had no such effect in the forenoon.

Experiment 226.—Successive records were taken of the effect of artificial darkness for two hours, alternating with exposure to light for two hours. The plant was subjected to darkness by placing a piece of black cloth over the glass cover from 12 to 2 p.m., it was exposed to light from 2 to 4 p.m. and darkened once more from 4 to 6 p.m.

The record given in figure 217 shows that the leaf had been moving upwards under the action of light (positive phototropism); darkness commenced at the point marked with a thick dot. *The after-effect on the stoppage of light is seen to be in the same direction as under light;* this persisted for ten minutes followed by recovery which was complete by 2 p.m., as seen in the horizontal character of the curve. On restoration of light (at the point marked with the second thick dot) the leaf moved upwards till the positive phototropic movement attained a maximum in the course of an hour and twenty minutes, after which neutralisation set in, and by 4 p.m. the positive phototropic effect had become partially neutralised. Arti-

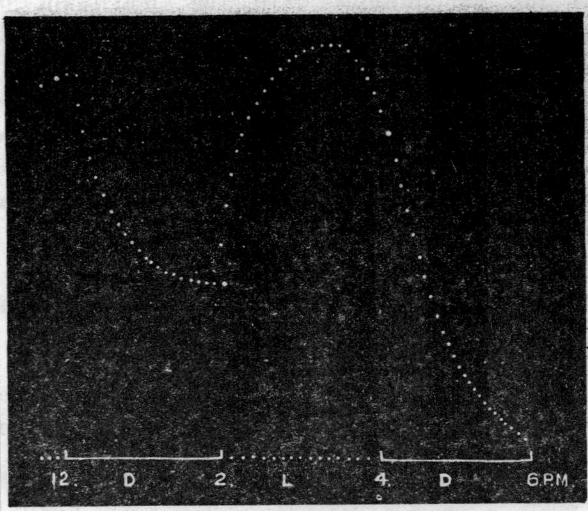

FIG. 217.—Effect of periodic alternation of light L, d of darkness D, on the response of *Mimosa* leaf. The first darkness causes the pre-maximal after-effect of slight erection followed by recovery. The subsequent application of light from 2 to 4 p.m. caused erectile movement followed by partial neutralisation by 4 p.m. Stoppage of light at the third thick dot caused a sudden fall of leaf *below* the position of equilibrium.

ficial darkness at the third thick dot caused a rapid down-movement which overshot the position of equilibrium. The difference of after-effect in the forenoon and in the afternoon lies in the fact that in the first case it was the pre-maximum after-effect; but in the second case the after-effect was post-maximum. I have already shown in the previous chapter that the pre-maximum after-effect of light is a short-lived movement in the same direction as under light, while post-maximum after-effect was a rapid over-shooting downwards beyond the equilibrium position. These characteristics are also found in the after-effects of light in *Mimosa*.

The responses of *Mimosa* on the cessation of light described above took place in the course of experiments which lasted for more than six hours. Objection may be raised that during this long period the temperature variation must have produced certain effects on the response In order to meet this difficulty, I carried out the following experiments which were completed in a relatively short time. I have already explained how the period of experiment could be shortened by suitable increase of the intensity of light. The experiment was commenced inside a room at noon and completed by 2 p.m.; the temperature variation during this period was less than 1°C.

After-effect at pre-maximum : Experiment 227.—Light from

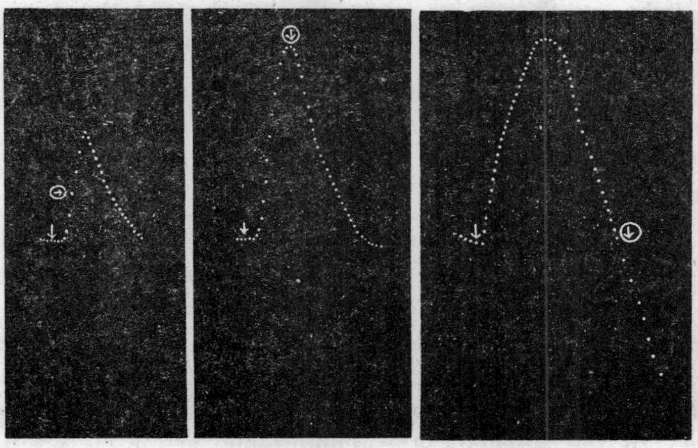

FIG. 218 FIG. 219. FIG. 220.

FIG. 218.—Pre-maximum after-effect of light in *Mimosa.*

FIG. 219.—After-effect at maximum.

FIG. 220.—Post-maximum after-effect exhibiting an 'over-shooting' below position of equilibrium.

In the above records light was applied at arrow, and stopped at the second arrow enclosed in a circle.

an 100 c.p. incandescent lamp was focussed on the upper half of the pulvinus of *Mimosa* for 8 minutes, after whic the light was turned off. The after-effect was a persistence of previous movement followed by recovery (Fig. 218).

After-effect at maximum: Experiment 228.—Continued action of light for 18 minutes induced maximum positive curvature as seen in the upper part of the curve becoming horizontal. On the stoppage of light, there was a recovery to the original position of equilibrium (Fig. 219).

After-effect at post-maximum: Experiment 229.—A fresh specimen of plant was taken for this experiment; it exhibited maximum positive curvature after an exposure of 20 minutes; continuation of light for a further period of 17 minutes produced complete neutralisation. Stoppage of light at this point, gave rise to a rapid down-movement (Fig. 220) below the equilibrium position.

The experiments that have been described show that the rapid fall of the leaf of *Mimosa* in the afternoon is due to 'over-shooting' which is the after-effect of prolonged action of light.

We are now in a position to give a full explanation of the different phases of diurnal movement of the leaf of *Mimosa*. The fall of the leaf commences from its highest position at thermal-dawn at 6 a.m. in the morning and continued till the thermal-noon at 2 p.m. This is the thermo-geotropic reaction due to rise of temperature In the forenoon the phototropic action is positive, and the fall of the leaf, due to rise of temperature, is brought about in opposition to the action of light. The temperature begins to fall after 2 p.m. and the leaf begins to erect itself, and in the absence of any disturbing factor would have continued its up-movement till next morning. But light undergoes a rapid diminution after 5 p.m. and the after-effect of light is an 'over-shooting' in a downward direction. This fall continues